Environments of Intelligence

"An absorbing volume that integrates an extraordinarily wide area of work, with interesting observations and new twists right to the end."

Ruth Millikan, University of Connecticut, USA

What is the role of the environment, and of the information it provides, in cognition? More specifically, may there be a role for certain artefacts to play in this context? These are questions that motivate "4E" theories of cognition (as being embodied, embedded, extended, enactive). In his take on that family of views, Hajo Greif first defends and refines a concept of information as primarily natural, environmentally embedded in character, which had been eclipsed by information-processing views of cognition. He continues with an inquiry into the cognitive bearing of some artefacts that are sometimes referred to as "intelligent environments". Without necessarily having much to do with Artificial Intelligence, such artefacts may ultimately modify our informational environments.

With respect to human cognition, the most notable effect of digital computers is not that they might be able, or become able, to think but that they alter the way we perceive, think and act.

Hajo Greif teaches at the Munich Center for Technology in Society (MCTS), Technical University of Munich, Germany, and the Department of Philosophy, University of Klagenfurt, Austria. His research interests cover the philosophy – and some of the history and the social studies – of science and technology, as well as the philosophy of mind.

History and Philosophy of Technoscience
Series Editor: Alfred Nordmann

For a full list of titles in this series, please visit www.routledge.com

1 **Error and Uncertainty in Scientific Practice**
Marcel Boumans, Giora Hon and Arthur C. Petersen (eds)

2 **Experiments in Practice**
Astrid Schwarz

3 **Philosophy, Computing and Information Science**
Ruth Hagengruber and Uwe Riss (eds)

4 **Spaceship Earth in the Environmental Age, 1960–1990**
Sabine Höhler

5 **The Future of Scientific Practice 'Bio-Techno-Logos'**
Marta Bertolaso (ed.)

6 **Scientists' Expertise as Performance: Between State and Society, 1860–1960**
Joris Vandendriessche, Evert Peeters and Kaat Wils (eds)

7 **Standardization in Measurement: Philosophical, Historical and Sociological Issues**
Oliver Schlaudt and Lara Huber (eds)

8 **The Mysterious Science of the Sea, 1775–1943**
Natascha Adamowsky

9 **Reasoning in Measurement**
Nicola Mößner and Alfred Nordmann (eds)

10 **Research Objects in their Technological Setting**
Bernadette Bensaude Vincent, Sacha Loeve, Alfred Nordmann and Astrid Schwarz (eds)

11 **Environments of Intelligence: From Natural Information to Artificial Interaction**
Hajo Greif

Environments of Intelligence

From natural information to artificial interaction

Hajo Greif

First published 2017
by Routledge
2 Park Square, Milton Park, Abingdon, Oxon OX14 4RN

and by Routledge
711 Third Avenue, New York, NY 10017

Routledge is an imprint of the Taylor & Francis Group, an informa business

British Library Cataloguing-in-Publication Data
A catalogue record for this book is available from the British Library

Library of Congress Cataloging-in-Publication Data
Names: Greif, Hajo, 1968– author.
Title: Environments of intelligence : from natural information to artificial interaction / Hajo Greif.
Description: New York : Routledge, 2017. | Series: History and philosophy of technoscience | Includes bibliographical references and index.
Identifiers: LCCN 2016049348 | ISBN 9781138222328 (hardback : alk. paper) | ISBN 9781315408101 (ebook)
Subjects: LCSH: Cognition. | Nature and nurture.
Classification: LCC BF311 .G74 2017 | DDC 153—dc23
LC record available at https://lccn.loc.gov/2016049348

ISBN: 978-1-138-22232-8 (hbk)
ISBN: 978-1-315-40810-1 (ebk)

Typeset in Times New Roman
by Apex CoVantage, LLC

Contents

Preface

This is a book about machines, intelligence and environments, but it is not a book about intelligent machines inhabiting some environment. Nor is it a book about machine intelligence as conceived of under the paradigm of Artificial Intelligence. Let alone is it a book about environments as intelligent agents of sorts. Instead, it is a book about human intelligence, how it is shaped by its environments and what role some kinds of artefacts play in that shaping. It sets out as an endeavour to defend and refine a concept of information as primarily *natural*, environmentally embedded information, which had been eclipsed by the notions of information processing that came to the fore with the rise of the computer. The argument continues with an inquiry into the cognitive bearing of some technologies that happen to be computational by design and are sometimes referred to as "intelligent environments" – and that do not necessarily have much to do with Artificial Intelligence. This work aims at a somewhat unconventional perspective on technologies whose relationship to human cognition may have been misrendered by the notion of intelligent machines and, vice versa, it aims at a somewhat unconventional perspective on human cognition whose nature may have been misrendered by models that rely on that same notion.

If one looks for a banner under which my project sails, that banner should read, in the naturalist spirit of a robust cognitive externalism: "Reasoning [. . .] is done in the world, not in one's head" (Millikan 1993b, 12). If reasoning commences from direct interaction with the world, it might be worthwhile to begin with taking a closer look at what that world does to and with our thinking rather than starting from a rather traditional, cognitivist or 'Cartesian' image of the mind in order to criticise it and then move on to claim that our thinking is situated in and extends into the world. (Anyway, that Cartesian image usually seems to serve as a convenient myth rather than being treated as a serious philosophical position worthy of serious critique.) The variety of externalism endorsed in what follows will be roughly in line with the views of extended, embedded, embodied and enactive cognition, also known by the family name "4E cognition". Even though not all of the following authors would agree to be subsumed under this rubric, and even though they would not all agree with my arguments, the closest match between what is presented here and other contemporary work in philosophy and cognitive science is, on my opinion, to be found in Chemero (2009), Clark (1997),

Godfrey-Smith (1996a) and Sterelny (2003), with Dretske (1981), Gibson (1979) and Millikan (1984) as the relevant proximate and the American naturalists as the historically more distant precursors.

Some readers may be struck by the lack of a clear distinction between the concepts of cognition and perception in what follows. However, there is a method to this apparent confusion: as some of the authors referred to in this book, and unlike the advocates of what is inelegantly called "cognitivism", and certainly unlike a great deal of philosophical tradition (which, I am told, was even more scrupulous in distinguishing among sensation, perception and cognition), I believe that there is no such clear distinction. There is no thinking without perceiving and acting in, and interacting with, an environment. Perception already *is* a form of interacting with the environment and, historically and systematically, all of the less action-bound activities of our nervous systems flow from such interaction.

In pursuit of its main argument, this book will incorporate some themes from the philosophy of mind, the philosophy of science (with focus on cognitive sciences and biology) and the philosophy of technology, but at its very heart it involves all three of them at once, and it draws on some disciplines outside philosophy. The specific constellation of disciplines involved is a reflection of my intellectual upbringing, but I am hopeful to make a credible case for the usefulness and non-arbitrariness of this combination – and perhaps for the virtue of a certain degree of intellectual generalism. If one still needs a rubric under which to file this work, that rubric may read "philosophy of mind-through-machines" (but certainly not conventional, qualia-and-supervenience philosophy of mind) or "philosophy of technologies of cognition" (but certainly not philosophy of engineering or critical theory of technology) or "philosophy of biology of cognitive functions" (but certainly not of a comprehensive sort), or perhaps to some extent "philosophy of some of the cognitive sciences" (but certainly not philosophy of Artificial Intelligence, and nothing near proficient in those sciences). Each of these labels will fit partly, none entirely, and all will look a bit awkward. Hence, I am aware of the risk of getting proverbially "caught between two stools" (or even more if that were topologically possible), but by virtue of incurring that risk, this book may have something to offer to several audiences.

So let me try to identify a few matches between parts of my argument and some communities to which these might be of interest: first, philosophers of mind interested in the varieties of naturalised semantics may find something worthwhile in my reconstruction of the notion of natural information and informational environments (see Chapters 2 through 6). Second, philosophers of perception or philosophers of psychology, and possibly some cognitive psychologists, might be interested in my reconstruction of what is ecological in the ecology of perception (see Chapter 3 in particular). Third, students of 4E cognition might see a contribution to their debates in my exposition of how cognition "extends" over and is "embedded" in the environment (see Chapter 7 in particular). Fourth, philosophers of technology and members of the science and technology studies community who are interested in digital technologies might find something to relate to both in the general outlook of the study and in the case studies presented towards the end of the book (see Chapters 8 and 9 in particular). And, fifth, computer scientists, designers and engineers working on smart environments might gather some useful information for their work from those same chapters. Reaching this latter audience I would consider the greatest success for this book, as it might stand a chance of, however indirectly and modestly, feeding into the shaping of those environments.

The unabashed naturalism of what follows has a serious epistemological implication that should be added as a disclaimer here: real-world scientific theories of perception, evolution and other phenomena as well as some of their empirical groundwork will be relevant and sometimes indispensable to my arguments. I hence not merely incur the risk of my arguments falling with the possible empirical falsification of the theories in question; I actually embrace that risk, happily embarking on the boat imagined by Quine (with the help of Neurath 1983, 92), in which philosophy and science sail together:

> A boat which [. . .] we can rebuild only at sea while staying afloat in it. There is no external vantage point, no first philosophy. All scientific findings, all scientific conjectures that are at present plausible, are therefore in my view as welcome for use in philosophy as elsewhere.
>
> (Quine 1969b, 126f)

Any serious quest for knowledge is exposed to the risks that come with this open-ended task.

Acknowledgements

Research on which this publication is based was funded by the Austrian Science Fund (FWF), grant number J 3448-G15 within the Erwin Schrödinger Programme.

The manuscript on which this book is based was first written over the course of nearly three years between October 2011 and August 2014, with several intermissions. Part of it was written when I was working at the Department of Science and Technology Studies at Alpen-Adria Universität Klagenfurt (AAU) – but only a small part of this part has been included in the present book. Most of what follows was researched and written, and everything was thoroughly rewritten, within the context of my FWF Erwin Schrödinger Fellowship at the Munich Center for Technology in Society (MCTS), Technical University of Munich (TUM). The full manuscript was then submitted as a thesis for "habilitation", that oddly Continental academic rite of passage for senior post-docs, with the Department of Philosophy at AAU, in late 2014. The manuscript has been heavily edited and revised since then, on the basis of the reviews I received and the second thoughts I gave to my initial ideas.

The contributions of those who supported and believed in my work, but also of those who criticised it, shall be duly acknowledged here, beginning with those of the reviewers of my thesis, Ursula Renz and Holger Lyre; those of Ruth Millikan, who had no obligation to read and pick apart central parts of the draft manuscript but generously did so anyway; those of my funding agency, FWF, and the anonymous reviewers who assessed the project proposal from which most of what follows grew; those of nine interview partners from the fields of computer science, robotics and ergonomics at TUM who helped me to ground at least some of my speculations about the nature of cognitive artefacts; those of the publishing house, Routledge, with Robert Langham and Michael Bourne to be named in particular, and the two anonymous reviewers they recruited for assessing the book proposal; those of Alfred Nordmann, the series editor, who first provided me with a chance of having my manuscript published with Routledge; those of several audiences at several conferences upon whom I tried out earlier versions of what follows; those of Arno Bammè, my former head of department in Klagenfurt, who first encouraged me to embark on this project; those of the Department of Philosophy at AAU, which took care of my habilitation project and proceedings after Arno's retirement; those of Klaus Mainzer, my host in Munich; and those of my other colleagues at MCTS. More detailed acknowledgements are to be found in the endnotes.

My greatest gratitude, however, is reserved for my wife, Karin, and our two children, David Marius and Simon Valentin. When I started writing, David was a tiny embryo enjoying the safety and warmth of his mother's womb. When I was in the thick of writing (or at least supposed to be so), Karin delivered him in good health. When I submitted the manuscript for review with the faculty, David could already walk and talk, play with other children and make fun of his parents, while Simon was just a few weeks away from being born. Now that proceedings, contracting and revisions are finished and the book is going to publication, Simon is steadily moving towards the same achievements. My accomplishments of the last few years invariably pale before those of my dearest ones.

Parts of this book are based on papers that have been presented at conferences, partly published in conference proceedings, and in one case as a peer-reviewed journal article:

- "What is the Extension of the Extended Mind?" *Synthese* Online First, 2015. Print version to appear in *Cognition* Special Issue edited by Cameron Buckner and Ellen Fridland. URL: http: //link.springer.com/article/10.1007/s11229-015-0799-9 (slightly extended version of Chapter 7).
- "Resurrecting Dretskean Information", submitted paper for the Ninth International Conference of the German Society for Analytic Philosophy (GAP.9), Osnabrück, Germany, September 2015 (short version of Chapter 2).
- "What is an Informational Environment?", submitted paper for the Sixth Workshop on the Philosophy of Information (WPI6), Society for the Philosophy of Information, Duke University, Durham, NC, USA, May 2014 (short version of Chapter 6).
- "Affording Illusions? Natural Information and the Problem of Misperception", invited paper for the interdisciplinary conference "What Affordance Affords", GOTO Project (Genesis and Ontology of Technoscientific Objects), Technische Universität Darmstadt, Germany, November 2013 (version of parts of Chapter 3).
- "Perceptual Illusions, Misperceptions, and the Empirical Strategy". In: *Mind, Language and Action. Papers of the 36th International Wittgenstein Symposium*. Ed. by D. Moyal-Sharrock et al. Kirchberg am Wechsel: ALWS, 2013, 163–165 (early version of parts of Chapter 3).
- "Wie denkt eine intelligente Umwelt? Modelle adaptiven Verhaltens in der Ambient Intelligence". In: *Vernetzung als soziales und technisches Paradigma*. Ed. by H. Greif and M. Werner. Wiesbaden: VS Research, 2012, 207–232 (early German version of parts of Chapter 9).
- "Erweiterte Phänotypen und Erweiterungen des Geistes: Die Rollen der Umwelt in der KI-Forschung". In: *Die Neukonstruktion des Menschen. Beiträge zum 7. Internationalen Tönnies-Symposium*. Ed. by A. Bammé. München: Profil, 2011 (early German incarnation of parts of Chapter 7).

- "Die Natürlichkeit künstlicher Intelligenzen und Umwelten", *Technikfolgenabschätzung – Theorie und Praxis* 20.1, 2011, 26–32 (early German incarnation of ideas related to Chapter 9).
- "Thinking with the Environment. Language, Pictures and Other Guides for the Extended Mind." In: *Image and Imaging in Philosophy, Science, and the Arts. Papers of the 33rd International Wittgenstein Symposium*. Ed. by R. Heinrich et al. Kirchberg am Wechsel: ALWS, 2010, 110–112 (early incarnation of parts of Chapter 8).
- "Where am I and Who are You? Coping with Ambient Intelligence", submitted paper for the Seventh European Computing and Philosophy Conference, Barcelona, Spain, July 2009 (earliest incarnation of ideas related to Chapter 9).

Permission for reusing a major portion of "What is the extension of the extended mind?", *Synthese*, Springer, 2015, for this book has been kindly granted by the publisher. This material is used as per the terms of the Creative Commons Attribution 4.0 International License (http://creativecommons.org/licenses/by/4.0/) and was slightly adapted by the author for inclusion in the present context. Permission to reuse portions of publications Greif (2010) and Greif (2013) for this book has been kindly granted by the Austrian Ludwig Wittgenstein Society.

This publication presents results of Austrian Science Fund (FWF) research grant no. J3448-G15. Open Access publication is supported by FWF publication grant no. PUB 488-Z24.

Der Wissenschaftsfonds.

1 Preliminaries

Ants and robots, parlour games and steam drills

> A man, viewed as a behaving system, is quite simple. The apparent complexity of his behaviour over time is largely a reflection of the complexity of the environment in which he finds himself [. . .] provided that we include in what we call man's environment the cocoon of information, stored in books and in long-term memory, that man spins about himself.
>
> (Simon 1969, 126f)

To some readers' surprise, perhaps, this thought was articulated by one of the pioneers of Artificial Intelligence (henceforth AI). Depending on emphasis, two markedly different readings can be given to it:

(1) One may focus on the human being as a behaving system, highlighting its relative simplicity while not further analysing the contributions from its environment. Under this reading, human cognitive achievements indeed appear rather modest.
(2) One may focus on the environment and the information it provides to the human being. Under this reading, environmental information is factored into the equation as a constitutive element of human cognitive achievements in all their richness.

Joseph Weizenbaum, an AI pioneer-turned-critic, bases his criticism of Herbert Simon's statement on the first of these two readings: such a statement, he contends, reveals a highly unfavourable view of human nature in which that nature is reduced to processes of a purely formal kind and to mechanically reactive behaviour (Weizenbaum 1976, 128–130). He does not even seem to consider the possibility of the second reading – which, however, appears to be the more plausible one in the context of Simon's argument. To be fair, it has to be noted that Simon first used the same phrasing to characterise the behaviour of ants (1969, 52), in order to provocatively substitute them with humans on the next page, and later adding human artefacts to the picture. The question is whether and to what extent the characterisation of the behaviour of ants as "quite simple" is supposed to be equivalent or analogous to that of human beings, with the normative implications that ensue.

The second of the aforementioned readings may seem somewhat untypical for what one expects from a pioneer of classical AI. Viewed in the light of that reading, Simon's hypothesis even appears to anticipate some of the core tenets of what has come to be labelled "active externalism" in the philosophy of mind, and the "4E" set of doctrines that flow from it, namely extended, embedded, enactive and embodied cognition.[1] Nonetheless, Simon remained committed to the notion of computational models of cognitive processes, as heralded by AI.

If we follow the second reading of Simon's hypothesis, human cognitive accomplishments do not appear unfairly depreciated but rich and complex in kind. That richness, however, depends first on the richness of the information available in human environments. Second, it depends on the richness of informational relations that human beings are actually in a position to exploit, in perception and action, within those environments. Only if these conditions are in place, human cognitive traits may also comprise representational capacities that finally, but still only partially, transcend the bounds of what the environment provides. This ranking is one of causal and, in terms of the evolution of cognition, historical priority. It does nothing to deny the importance of the 'internal' component of cognition, but makes the richness that there is to the mental life of human beings dependent on factors that are not in our heads.

In this book, I will try to develop a perspective on how Simon's insight into the cognitive importance of the environment and the information provided by that environment could be accounted for in general, and what implications this analysis will have for an account of information-providing artefacts like the ones referred to in the earlier quote. In doing so, I will have made a statement on the accomplishments of AI and the cognitive sciences with respect to what are important components of cognitive processes. However, that statement, unlike the Weizenbaum – Searle – Dreyfus continuum of fundamental critiques of AI, neither dares to dictate solutions to the problems that AI and the cognitive sciences have to deal with nor lectures them about the ultimate futility of their endeavours. In order to provide the reader with an idea of what follows in the two parts of this book, I will first sketch the backdrop against which I develop my approach, and then outline my argument.

Background

> The new form of the problem can be described in terms of a game which we call the 'imitation game.' It is played with three people, a man (A), a woman (B), and an interrogator (C) who may be of either sex. The interrogator stays in a room apart front the other two. The object of the game for the interrogator is to determine which of the other two is the man and which is the woman. [. . .]
>
> We now ask the question, 'What will happen when a machine takes the part of A in this game?' Will the interrogator decide wrongly as often when the game is played like this as he does when the game is played between a man and a woman? These questions replace our original, 'Can machines think?' [. . .]
>
> The new problem has the advantage of drawing a fairly sharp line between the physical and the intellectual capacities of a man.
>
> (Turing 1950, 433f)

Or is it an advantage? Both AI and its philosophical critiques to a large extent have been and remain living under the spell of the notion of the Turing Test, which essentially depends on drawing that "fairly sharp line". Taking this line to be a firm boundary rather than a heuristic, that test evolved into something quite far removed from Alan Turing's classical "imitation game", as described in the previous quote. It is the notion of such a firm boundary that has attracted criticism from the camp of 4E cognition with some justification and in some detail.

However, as has been argued with some justification, too (most prominently by Copeland 2000; Moor 1976; Whitby 1996), the aim of Turing's thought-experimental game of machines imitating human conversational behaviour was *not* to prove that machines could think, or to provide a test for whether they can think. After all, the "imitation game" was playfully fashioned after an eponymous popular Victorian parlour game.[2] Instead, that game was part of Turing's inquiry into the possible scope and depth of the tasks he designed for a certain class of theoretical machines when devising his theory of computability in Turing (1936). He considered digital computers one possible material incarnation of those theoretical machines – and human computers another, which served as the blueprint for the former.[3]

Turing's theory of computability was driven by a genuinely meta-mathematical interest: if, within the confines of a logical calculus, there is an unequivocal, well-defined and finite, hence at least in principle executable procedure for deciding on the provability of a proposition that has been stated in that calculus, this procedure should be translatable into arithmetical forms. These, in turn, could be broken down into a set of simple mathematical routines that would be executable for a human 'computer', that is a person with basic mathematical skills who is provided with a set of input numbers and a set of instructions and then ordered to calculate the result (and pass on that result as input for further computation). These inputs and instructions, Turing's argument continues, would be simple and unequivocal enough to be handled by a suitably designed machine, too.

Although representing only an expressly restricted subset of human intellectual abilities, these tasks were well beyond the scope of traditional, cog-belt-and-pulley machines. Margaret Boden (2006, 168f) notices that the thought that machines could possibly think was not even a 'heresy' up to the early twentieth century, as that claim would have been all but incomprehensible. The reason is that intellectual capabilities were thought of as essentially tied to the (most likely) human organism whereas the capabilities of machines were mostly thought of as being restricted to the exertion and harnessing of physical forces. More precisely, according to machine-age definitions of machines, "the primary function of a mechanical device can be either the modification of motion (direction) or the modification of motion and force (amplification and reduction)", where the former would be a mechanism and the latter a machine proper (Mitcham 1994, 170, in his reconstruction of definitions of machines that go back to the mid-nineteenth century). A more reductive definition confines the activities of machines to "changing the direction of motion" of matter, because "moving matter is all the force with which machines deal" ("What is a Machine?" 1872, 39). A more differentiated classic definition conceives of a machine as "'a closed kinematic chain' or 'a combination of resistant bodies so arranged that by their means the mechanical forces of nature can be compelled to do work accompanied by certain determinate motions'" (Mitcham 1994, 170f, quoting Franz Reuleaux's 1876 book *Kinematics of Machinery*). In seeking to extend the domain of what machines can do beyond kinematics (and, by extension, the effects of electrical energy), Turing aligned himself with a small tradition of thinkers and inventors who conceived of machines that could serve logico-mathematical purposes beyond basic calculating functions, most notably Charles Babbage (1864) and Gottfried Wilhelm Leibniz (1685). In abstraction from what cog-belt-and-pulley machines do, a machine was now conceived of as any arrangement of discrete elements and the discrete operational steps associated with them which, with determinate regularity, transforms a certain input into a certain output.

With respect to comparing human and machine accomplishments on the basis of this theoretical concept, the key novelty introduced by Turing's theory of computability was twofold: first, his theoretical machines, in analogy to human computers, could switch between different sets of instructions, and hence between different logico-mathematical tasks, without necessitating a change to the structure of the machine proper. The input-output relations and a certain subset of the machine's operations were now conceived of as modifiable. Second, the same type of logical operations could be realised in a variety of different systems, from human beings to specific machines, in a variety of ways, and thus in abstraction from their physical characteristics. The first part of this twofold idea is entailed by Turing's description of his theoretical "Universal Machines", whereas the second has become known as "Turing Machine functionalism". It includes opening up the conceptual possibility of artefacts to accomplish tasks that would count as thinking in a human. However, this is an implication (not a corollary) of the larger argument, which is not about human beings or machines in particular but about a meta-mathematical method. Above all, the second part of Turing's idea waives any requirement that those accomplishments would have to be reached by the same means in the machine analogue.

Hence, a superficial reading of the imitation game is prone to misguiding the reader into seeking for an analogy between human and machine accomplishments that is closer than intended. In one of the first philosophical essays that took Turing's thought experiment seriously, Keith Gunderson (1964) captures a point that seems to have been lost in the long-lasting, and arguably not always fruitful, philosophical debates around the possibilities and limitations of AI:

> One might well contend that machines can't think, for they do much better than that. [. . .] Machines can almost instantaneously and infallibly produce accurate and sometimes original answers to many complex and difficult mathematical problems with which they are presented. They do not need to "think out" the answers. In the end the steam drill outlasted John Henry as a digger of railway tunnels, but that didn't prove the machine had muscles; it proved that muscles were not needed for digging railway tunnels.
>
> (Gunderson 1964, 244f)

Competing with the steam drill in an epic battle at the height of the industrial age, our folk ballad's hero scored the costliest of victories. Exhausted from his successful race against the machine, John Henry was found "dying with his hammer in his hand". In this tragic manner, he has become an African American folk hero, as a symbol of black workers' struggle for respect in a society where many odds were against them. One of these odds was the progressive displacement of their labour by machines.

In keeping with Turing Machine functionalism, the previously quoted passage suggests that such displacement normally works precisely because machines do *not* reproduce or replicate human organic features and behavioural patterns, but because they achieve the results of goal-directed human behaviours more quickly, more effectively and more efficiently – or because they enable the achievement of certain, hitherto unattainable, goals in the first place. The steam drill outlasted John Henry not because it was like him, only stronger, more enduring, but because it was very much *un*like him. Hence, analogy in function does not require similarity in structure. Provided that the task in question is well-defined, any structure that produces effects that result in the accomplishment of that task, and that produces these effects in accordance with an equally well-defined regularity (rather than coincidentally), will count as functionally equivalent.

Turing's specific concern with logico-mathematical tasks aside, his style of functionalist argument has a long pedigree in biology, where analogies in the functions of traits can be determined for phylogenetically unrelated species, whereas sameness of kind depends on common descent of some trait – and only on common descent. The underlying physiological structure might remain identical over the course of generations, although both the function which it serves and its observable shape may be subject to change if conditions to which that trait responds are altered. The independence of functional from structural relations in biology was captured in the conceptual distinction between "analogues" and "homologues" by Richard Owen (1848). This distinction was put on an evolutionary footing by Charles Darwin, who provided vivid examples of homology himself when he asked:

> What can be more curious than that the hand of a man, formed for grasping, that of a mole for digging, the leg of the horse, the paddle of the porpoise, and the wing of the bat, should all be constructed on the same pattern, and should include the same bones, in the same relative positions?
>
> (Darwin 1859, 434)

Hence, on the one hand, one ancestral form, for example a mammal's forelegs, may come to serve divergent functions – as wings in bats and arms in human beings. On the other hand, one and the same function may come to be accomplished by structures of various degrees of similarity that have developed along independent pathways in genealogically remote species – as for the structurally similar but evolutionary independent lensed eyes in vertebrates and many Cephalopods, and the structurally dissimilar and equally independent compound eyes in Arthropoda.

However, this evolutionary type of functionalist argument cuts both ways: first, it exposes a fallacy in the claim made by AI critics that a disanalogy in structure, that is in the way in which and the means by which some task is accomplished, precludes analogy in function. This is basically the inverse of Turing Machine functionalism, or indeed of any functionalist argument of this kind, and it appears to be the bottom line of John Searle's AI critique in (1980) and many other critiques that follow his route (which include philosophers as eminent as Dretske 1985 and Fodor 1981). If similarity in structure were thus required, and unless that similarity were superficial or coincidental, the requirement would be one of homology. This is a condition that, by definition, any machine would invariably fail at. Moreover, it is a condition that does not respond to the claim of functional analogy, as homology and analogy are independent affairs. Second, and for the same reason, similarity in structure is uninformative as to an analogy of function. Hence, human-likeness of some machine in appearance or behaviour does not warrant any inference with respect to intellectual abilities of that machine.

Despite this implication of a functionalist argument, and in keeping with the questionable interpretation of Turing's imitation game as a test for, or definition of, intelligence, AI was long committed to the criterion of human-likeness of its systems, in appearance and behaviour. This commitment is reflected in the definition of the research programme that kept dominating the field at least until the 1990s: AI was understood as an inquiry into the nature of the human mind in which theories about its structure, properties and functions were tested by means of computer programmes, computer systems or robots as models. However, the long-standing dominance of AI as the modelling and simulation of human thought processes may have helped to obscure rather than illuminate other, possibly more instructive, roles of computers and robots as models concerning (if not to say *of*) human cognition and action.

Turing himself deliberately styled his inquiries into the possibility of thinking machines as flights of fancy that he only could – and did – wish to be true, but he also pointed towards two implications of the presence and use of computing machinery that is much closer to home in many respects:

> The original question, 'Can machines think?' I believe to be too meaningless to deserve discussion. Nevertheless I believe that at the end of the century the use of words and general educated opinion will have altered so much that one will be able to speak of machines thinking without expecting to be contradicted.
>
> (Turing 1950, 442)

Irrespective of the question of whether they match human intellectual abilities, the accomplishments of computing machinery would ultimately *alter our understanding of what human thinking is* – a point made so explicit by Turing that I keep wondering why it has not raised more scholarly attention. Digital computers may be relevant to human cognition, and to how we conceive of it, in other ways than simulating human thought processes.

As the flip side of the same coin, the accomplishments of computing machinery will also *alter our understanding of what a machine is*. Given that advanced tool use is a uniquely human characteristic, a change in what artefacts can accomplish, and a change in human perception of those accomplishments, will have an equally pivotal influence on the human condition. Both changes that Turing had in mind are happening right before our eyes.

The paradigm of what I have in mind here has been articulated, in related but slightly different fashion, by two computer scientists who wrote half a century after Turing inaugurated that very discipline in his 1936 paper: Rodney Brooks and Mark Weiser. Both authors highlight the importance of the relations to the respective environments of the systems they were developing while relegating those intellectual accomplishments which were central to classical AI to secondary importance.

As one of the first proponents of "Nouvelle AI", Brooks argues that traditional AI systems were incapable of grasping important aspects of human cognition because these aspects are not located in human beings' heads but in the environments in which they act and interact. Thus, inclusion of these environmental aspects in embodied, that is to say robotic, systems has to take precedence over criteria of human-likeness or mental representation. It is best accomplished in bottom-up and modular fashion, starting from fairly simple systems, Brooks continues:

> It seemed a reasonable requirement that intelligence be reactive to dynamic aspects of the environment, that a mobile robot operate on time scales similar to those of animals and humans, and that intelligence be able to generate robust behavior in the face of uncertain sensors, an unpredictable environment, and a changing world. [. . .] Internal world models that are complete representations of the external environment, besides being impossible to obtain, are not at all necessary for agents to act in a competent manner. Many of the actions of an agent are quite separable – coherent intelligence can emerge from independent subcomponents interacting in the world.
>
> (Brooks 1991, 1228)

No AI system will capture the mechanisms responsible for the specific patterns of organism-environment interaction unless it is able to capture their purpose, too, which, in turn, can be accommodated only by modelling the organism-environment interaction in a most direct way. Directness in this sense does not require the creation of similes of traits or behaviours on a phenomenal level, nor will it suffice to consider the structures and processes inside the organism. Instead, one will have to identify, and factor into the equation, specific couplings between variables within organism and environment, and the emerging patterns of interaction between them. This is the premise on which the research programme of "behaviour-based AI" was developed (for statements of this programme, see Beer 1995; Brooks 1999; Maes 1993; Steels and Brooks 1995).

Weiser's work, in turn, is considered visionary in setting the stage for, and coining the term of "ubiquitous computing" (to which he also referred as "embodied virtuality"). He describes his ideas in hardly less vivid terms than those used by Turing to illustrate his concept of computing machines:

> In our experimental embodied virtuality, doors open only to the right badge wearer, rooms greet people by name, telephone calls can be automatically forwarded to wherever the recipient may be, receptionists actually know where people are, computer terminals retrieve the preferences of whoever is sitting at them, and appointment diaries write themselves. No revolution in artificial intelligence is needed, merely computers embedded in the everyday world. [. . .]
>
> Machines that fit the human environment instead of forcing humans to enter theirs will make using a computer as refreshing as taking a walk in the woods.
>
> (Weiser 1991, 99 and 104)

Under this paradigm, human thinking is primarily considered with respect to practical tasks in everyday contexts, and as being mostly implicit in nature. Computer systems embedded in the environment may unobtrusively assist in such practical, implicit problem solving. Reference to AI is even explicitly avoided. Despite these apparently humble aims, we encounter machines that accomplish things that were not anywhere near conceivable as machine accomplishments in Turing's day. (Arguably, even Turing himself had other accomplishments in mind.)

Although AI remains a factor in computer science and engineering, and although computers have not 'disappeared' into the background of the environment altogether since Weiser wrote the aforementioned lines, this shift in emphasis is significant. As evidenced by the attention that Brooks's and Weiser's works have received, and continue to receive, this shift in emphasis affects the development both of the cognitive sciences, with AI as one of their founding disciplines, and of user-oriented computing applications. On either level, human cognitive abilities will appear in a different light when viewed from the perspective of the environments in which they develop and unfold.

Outline

On the background of these preliminary observations, the aim of this book can be parsed into two related questions, which roughly correspond to the two parts of the text: first, what is the role of the environment, and of the information it provides, in cognition? Second, and more specifically, what role do artefacts play in this context? Part of my mission will be to demonstrate that the focus on artefacts in the second question is not an arbitrary narrowing down of the topic but inherent to the first question – at least as far as *human* cognition is concerned – and that this condition continues to hold if I further narrow down my focus on specific kinds of artefacts.

Both in philosophy and in the cognitive sciences, it is a relatively recent but now well-established proposition to seriously consider a constitutive role of an organism's environment to his cognitive abilities and their accomplishment. So, as to the first question, I will present a take on the role of the environment in cognition that shares the notion of such a constitutive role but claims distinctiveness in how it conceives of the information encountered in the environment. More precisely, I will seek to rehabilitate the notion of natural information introduced by Fred Dretske (1981) against the altogether reasonable criticisms brought forward by advocates of 4E cognition who, following Ruth Millikan (2001), argue for a local and probabilistic rather than nomologically governed character of natural information. In fact, these criticisms will be helpful in carving out the main point to be defended here. In the service of that defence, I will align Dretskean information with a reading of the notion of information for perception in the ecological approach to perception (Gibson 1979).

As to the second question, both sides involved in the aforementioned debates either confine their considerations of the role of the environment in cognition to the *natural* environment or they have a restricted view of the role and importance of artefacts in human environments. However, throughout all of human history and much of prehistory, artefacts have been invented and used that remain deeply involved in how human beings perceive and act in their environments. One may think of carvings, figurines, pictures and inscriptions back then or books, broadcast media, phones and computers now. Basically, one may think of all things manufactured that do not primarily aim at altering physical conditions in the environment but at providing information in some way. Their function is to augment, supplement, substitute or, in some cases, even first enable, human cognitive functions. It will be helpful to conceptually distinguish between these types of contributions by artefacts, which I will henceforth refer to as "cognitive artefacts". If one considers language an artefact, it will be the paradigm of strongly constitutive cognitive artefacts.

My main focus in this context, however, will be on a subset of cognitive artefacts whose very existence owes to Turing's work, namely digital computers. Among these, I will focus on computer-based artefacts that are directly concerned with their users' perception of, and interaction with, their environments. I will hence seek to decouple my account of those artefacts from the often poorly specified criterion of human-likeness that holds sway over much of what followed upon classical AI – including parts of 4E cognition. I will try to demonstrate that, with respect to human cognition, the most notable effect of digital computers and their manifold descendants is not that they might be able, or become able, to think or to provide a simile thereof, but that they alter the way we perceive, think and act. What they do is to modify our informational environments. Whereas all technologies have the effect of materially changing human environments and perhaps also of affecting our perception of and interactions with these environments, those technologies which directly operate on information have a capacity of equally directly modifying the patterns of available informational relations by which we steer. Despite the involvement of artefacts, and despite being technologically modified, the information involved shall remain being conceived of as natural information.

Changes in the informational environment, I will continue to argue, prompt acts of accommodating or "naturalising" them. These effects of cognitive artefacts also work to question the boundary between what is commonly considered the inner workings of the mind and the role of the outer environment in cognitive processes. The technologies in question do not even have to be extremely sophisticated or, in the conventional sense, intelligent, to achieve these effects. The way in which they make available, possibly modify and present the information on which human action depends will make the difference. The incorporation of simulated or virtual elements in human environments will be the paradigm of the changes I have in mind but not their only manifestation. Typically, the purpose of such elements of "smart environments" is to enable quasi-natural interactions between human being and machine in a given context. Under some circumstances, the machine need not even be a – machine- or human-like – counterpart that could be easily identified or individuated as such by its human partner. Although not literally, a thus-designed artefact might cease to be a machine in a relevant sense.

In order to develop this argument, I will use Part I of this book to present an account of natural information, as encountered in our environments, and its contribution to "the apparent complexity" of our behaviours as referred to by Simon – which is not merely a seeming complexity, but that complexity which we can observe in and of ourselves. After all, what Weizenbaum still held to be an "impoverished" conception of man might ultimately be the richer one. I argue that information and environments are generally objective affairs but that organisms (human and other) live in informational environments that are, in a qualified sense, particular to them and that are constituted by the specifically relevant variables that they have to keep track of in order to get by. I will use Chapter 2 to spell out the notions of natural information and its role in perception and behaviour on a general level, with focus on the Dretskean view. In Chapter 3, I will discuss three theories of visual perception that give markedly variant importance to the role of information and of the environment of perception and action: David Marr's computational theory of vision, James Jerome Gibson's ecological psychology and Dale Purves's Empirical Strategy of vision. This discussion, and in particular a comparison between Gibson's and Dretske's concepts of information, will help me to carve out a notion of natural information in more detail in Chapter 4, and it will prepare the ground for a hypothesis on informational environments in Chapter 6.

In Part II, I will match my account of informational environments against what may seem its closest relatives in contemporary debates, namely the 4E family of theories, so as to sort out relevant commonalities and differences and prepare the ground for an account of cognitive artefacts (Chapter 7). I will then develop that account of cognitive artefacts in more detail (Chapter 8). In order to ground that still fairly theoretical discussion, I will proceed with brief inquiries into some technological attempts at taking the role of the environment in cognition seriously (Chapter 9). None of these technologies aim at modelling, simulating or explaining the inner workings of the human mind, but they may help to practically, and productively, undermine the very distinction between these 'inner workings' and their 'outer' environment, and they are likely to make a difference to how our informational environments stand. In conclusion, I will offer an outlook on the wider implications of my account on philosophical issues (Chapter 10).

In focusing on the role of the environment rather than the inner workings of mind, I will appear to skirt the long-standing issues of qualia and propositional content in the philosophy of mind. However, rather than merely cheating my way around or flatly denying the importance of phenomenal qualities and the intentionality of the mental, I will seek to make a case for the relevance of what is in the environment, and our interactions with it, to the existence and richness of these phenomena.

Notes

1 The term "active externalism" was coined by Clark and Chalmers (1998), but see also Hurley (1998), which was published together with Clark and Chalmers's more famous paper and provides a much more systematic account of externalist positions, and Wilson (1994). The moniker "4E cognition" was later established as a generic term for the class of extended, embedded, enactive and embodied cognition by Menary (2010c) for the *Phenomenology and the Cognitive Sciences* Special Issue "4E Cognition: Embodied, Embedded, Enacted, Extended" which he edited.

2 It is interesting to observe that this possible historical source is rarely acknowledged (but see Evans and Collins 2010, 60). Most scholars either take the game to be a proper scientific test, whereas others focus on the gender rather than the game aspect of that game; see, for example Saygin et al. (2000) and Sterrett (2000).

3 This direction of modelling from human being to machine is explicitly stated by Turing (1950, 436) himself and highlighted in the reconstruction of Turing computability in Copeland (2009). Also compare Ludwig Wittgenstein's (1984, 197) observation: "Turings 'Maschinen'. Diese Maschinen sind ja die *Menschen*, welche kalkulieren." ("Turing's 'Machines'. These machines are *humans* who calculate.") If there were a real machine involved in the design of Turing's machines, it was the mechanical typewriter, suggests Turing's biographer, Andrew Hodges (1983, 96–98).

Part I

Informational environments

2 Resurrecting Dretskean information

When choosing the title for this chapter, I considered adding some qualifiers to it, such as resurrecting Dretskean information "in part" or "(somewhat)". I ultimately chose to stick with something more clear and crisp though. One of the purposes of this chapter is to demonstrate that there are some insights to Fred Dretske's theory of meaning in his *Knowledge and the Flow of Information* that are worth defending, and that they have to do with his concept of information. Another purpose of this chapter is to elucidate the distinction between the concept of information defended here and the one established in the mathematical theory of communication (Shannon and Weaver 1949), and to highlight the importance of this distinction to what follows.

This chapter will begin with a brief outline of the notion of information and its role in an organism's behaviour on a general, introductory level. I will then dig more deeply into the issue of natural information, trying to recover the main insights of Dretske's account of natural information and discussing some of its difficulties. The chapter will conclude with a consideration of some implications of this view with respect to the phenomenon of intentionality.

Information, behaviour and probability

> The frog does not seem to see or, at any rate, is not concerned with the detail of stationary parts of the world around him. He will starve to death surrounded by food if it is not moving. His choice of food is determined only by size and movement. He will leap to capture any object the size of an insect or worm, providing it moves like one. He can be fooled easily not only by a bit of dangled meat but by any moving small object. His sex life is conducted by sound and touch. His choice of paths in escaping enemies does not seem to be governed by anything more devious than leaping to where it is darker. Since he is equally at home in water and on land, why should it matter where he lights after jumping or what particular direction he takes? He does remember a moving thing providing it stays within his field of vision and he is not distracted.
>
> (Lettvin et al. 1968, 234)

According to these observations, the frog lives in a fairly simple environment. To the frog, objects that do not move are just like objects that do not exist. Those objects which move in a certain way, and which are also of a certain size, will indiscriminately be treated as food. And the frog's anti-predator strategies seem to be exhausted by quickly moving away from illuminated areas. If a behaviourist psychologist had to dream up an organism to fit his theory, where pure stimulus-response mechanisms reign supreme and qualitative mental states either do not count or are denied to exist to begin with, this would be it – at least on Lettvin et al.'s account. However, the frog's apparently rather modest accomplishments have done nothing to diminish the lifespan, in evolutionary terms, of the frog as a species, although they might have contributed to many an individual frog's life being cut short by a more clever animal that professes in preying on frogs.

The general idea to be promoted by this example is that, on the one hand, human environments are certainly more complex than frog environments, despite being governed by the same laws of physics, and despite human beings and frogs inhabiting much of the same segments of space-time. On the other hand, one general condition applies to frogs and human beings alike. This condition is neatly captured by Dretske:

> There is no difference [. . .] between what *happens to* an electron in a magnetic field and what an electron *does* in a magnetic field. There definitely *is* a difference between what happens to an animal placed in water and what it does when placed in water.
>
> (Dretske 1988, 11, emphasis in original)

Different things will happen to different animals when placed in water. Some may drown, whereas others will depend on that very placement for survival, whereas frogs will be indifferent, under many circumstances, to being placed on land vs. in water. Quite obviously, water will be a different thing in many ways to different animals. However, although different animals will do different things when placed in water, the general kind of exchange is common to all animals (and perhaps in some respects to plants, too). That exchange is behavioural and informational in kind, with these two aspects being closely intertwined. In conjunction, these aspects set an organism's exchange with his environment apart from the exchange between, say, a pebble or a spoonful of salt and the water in which they are placed.

As to the behavioural aspect of the exchange between organism and environment, it is obvious, on the one hand, that identical physical conditions, such as heat or pressure, will have the same determinate effects on all organic matter as such, just as magnetic fields will have the same determinate effect on electrons. For example all organic matter of the kind that can be found in organisms will be subject to processes of pyrolysis or combustion at elevated temperatures; blood and other body liquids will invariably start to boil at body temperature when exposed to an atmospheric pressure of 6.3kPa; no organic matter will ignite when placed in water. On the other hand, however, changes in physical variables in different organisms' surroundings, unless they affect the composition and integrity of organic matter as such, may trigger or structure behaviours by means of which various organisms react to these changes. Such changes in the environment include the waning of daylight for plants, diurnal and nocturnal animals or the overall reduction of daylight hours and ambient temperatures over some weeks for migrating birds and hibernating mammals. Unlike the effects of physical forces on inert matter, a behaviour is not exhausted by the observable effects of such forces, as for example impact or heating, on organisms. Behaviour is a structured response to a condition or a set of conditions in the environment, and it is internally caused by the organism. This is the distinction Dretske was after.

As to the informational aspect of the exchange between organism and environment, all behaviours, and many things that are not immediately discernible as behaviours, are structured by informational relations. It is not the physical nature of the conditions in the environment as such that determines the response. Instead, a certain change of variables, or a certain constellation of constancies and changes of variables in the organism's surroundings, has a signalling effect to the organism. It is the presence or absence of a signal or the difference between signals that accounts for a difference in behavioural responses. Instances of some environmental variable moving outside a certain neutral or equilibrial range will, by virtue of the value assumed by the variable rather than by virtue of the electrical, chemical or other properties that are hence subject to variance, prompt a negative feedback reaction that results in returning the value of the variable into the equilibrial range.[1] This is the first sense in which informational relations are partly independent of, or underdetermined by, physical conditions.

The second sense in which informational relations are underdetermined by physical conditions is complementary to the first: as the functional and physical structure and dispositions of the organism are crucial structuring features in the generation of behaviour, a condition identical in physical terms may give rise to widely variant behaviours, depending on the informational relations in which it is embedded. Hence, frogs and humans will do different things when placed in water. Being so placed conveys different information to frogs (to whom it is the normal habitat) and to people in different contexts (Am I going swimming, or did I slip when walking on the river bank?). I will argue in Chapter 6 that, in providing different organisms with information for different behaviours, the set of variables that an organism responds to in his specific way constitutes his specific environment.

As the flip side, as it were, of the coin of underdetermination, information is to be thought of as an objective commodity, despite being used in diverse ways, in diverse constellations and to diverse purposes by different receivers or "consumers", for being relevant to them in distinctive ways. Information does not reduce to whatever an interpreter takes to be information. In some important respects, it is even indifferent to any interpreter's dispositions or the presence of any interpretation. Instead, it is a relation between world affairs that can be both independently observed and subject to probabilistic and context-specific modes of analysis. This view of information will be defended here as the proper reconstruction of Dretske's theory.

This view of information might be best understood when properly placed among the broad variety of theories of information that exist in the literature. In the first instance, it sees itself as part of a historical lineage of approaches that departed in some way from the mathematical theory of communication, as originally proposed by Claude Shannon and Warren Weaver (1949). That theory was not concerned with natural information and its role in animal cognition but with technical requirements for reliable signalling in telecommunications. Although one might argue that it is thus wholly irrelevant to the explanatory purpose at hand, it has continued to influence the debate for many decades and remains a common reference point for all sorts of theories of information, including the semantic concept of information introduced by Yehoshua Bar-Hillel and Rudolf Carnap (1952) and refurbished by Luciano Floridi (2004). Even in theories that are far removed from Shannon and Weaver's original purposes in grounding theories of meaning in a notion of natural information, the mathematical theory of communication remains a reference point. This influence is most prominent in Dretske (1981) but has proven somewhat difficult to parse and has been countered with accounts that partly or wholly distance themselves from the mathematical paradigm (e.g. Millikan 2001; Skyrms 2010; more on this debate in the following section).[2]

The mathematical theory of information, at the time of its introduction, was duly concerned with genuine engineering problems in telecommunications, namely with the requirements for reliable and effective transmission of messages between sender and receiver – signal rates, channel capacities, measures of levels of the noise that interferes with signal transmission and the degrees of redundancy that could compensate for noise-induced loss (Weaver 1949). However, there is a core set of tenets of the mathematical theory that can be formulated without reference to senders and receivers of signals as intentional communicators of messages – and arguably also without implicitly presupposing their presence. The minimal formal characteristics of information can be summarised in the following two definitions:

(IN-1) Any sequence of events in which a set of *possible* states of affairs at the source $\{s_1, \ldots, s_n\}$, as determined at t_0, is reduced by that state s_1 within the set which turns out to be the *actual* state at t_1, is an instance of information. The amount of information x involved is measured as the logarithmic function $x = log_2 \frac{1}{p}$ from the probabilities p of antecedent states to the actual state.[3]

In a second step, the conditions for the transmission of information will be accounted for, so that the relation between the transformations of probabilities at the source and those at the receiving end is determined:

(IN-2) The reduction of possibilities at the source must be sufficient to reduce the uncertainty on the receiving side R under conditions of "lossy" transmission.

Transmission is lossy

(a) if information produced at s is lost in *equivocation*, or
(b) if *noise* added to r compromises the information received at R.
Information is successfully transmitted only if an amount of bits is received that reduces the possibilities to an extent that is sufficient for specifying R's response.

Hence, at least as much additional or "redundant" information must be produced at the source as will have been cancelled out by the intervening noise and equivocation. If for example 50% of the information produced at the source is lost in transmission or if 50% of what is received at *R* is, in fact, not produced at the source, and if the losses are randomly distributed over the set of signals under consideration, the number of bits will at least have to double.

To begin with, IN-1 says that there must be some reduction of possibilities at the source *s* in order for information to obtain (a condition that will receive some qualification in the first section of Chapter 4). For example if there are eight equiprobable antecedent states at the source, it takes three binary decisions and thus three bits of information to attain a state of certainty. For sixteen equiprobable states, the required amount of information will be four bits. If the set of possible states determined at t_0 composes only the actual state that will be observed at t_1, and thus if the probability of s_1 to obtain is 1, *no* information is generated. Entirely static conditions at the source, with no intervening factors that could produce an altered state of affairs, would be a case in point. If, in contrast, a set of (roughly) equally possible states at the source at t_0 approaches infinity, and hence if the actual state s_1 at t_1 is extremely improbable to obtain, the realisation of s_1 is equally rich in information.

On this analysis, it appears that, to quote Weaver (1949, 14) again, "the words information and uncertainty find themselves partners". Prima facie, uncertainty seems to be related to unwelcome phenomena like noise that interferes with communication, and it raises a semantically relevant problem that was first identified by Bar-Hillel and Carnap (1952) and hence christened the "Bar-Hillel Carnap Paradox" (see Floridi 2004): the authors observed that, on a formal analysis, a self-contradictory sentence carries the highest amount of information, so that "it is too informative to be true" (Bar-Hillel and Carnap 1952, 8). In the face of such unwelcome implications, Weaver highlights the distinction between what he calls "undesirable" and "desirable" uncertainty (1949, 12f). The latter is to be found in the freedom of choice at the information source in selecting a message from a set of possible messages. Hence, a higher degree of freedom is correlated with a higher degree of uncertainty. If no alternative possibilities to some message exist at the source, there will be no information. As the alternative possibilities multiply, so does the amount of (desirable) uncertainty – and hence the amount of information that is available.

In thus distinguishing between uncertainty – and hence information – that is desirable vs. undesirable to sender or receiver, and in referring to the degrees of freedom at the source in selecting a message, it is implied that the information in question has to be sent by, or to originate in relation to, an intentional agent, that is a sender with a purpose. Only a sender with a purpose will be capable of goal-directed action that allows for some degree of deliberation as to what goals shall be attended by what means and on what grounds. Only such a sender will be in a position to *select* a message from a variety of alternative possibilities.

There seems to be an ironic twist in the positions adopted by the mathematical and the naturalistically inclined semantic theories of information: on the one hand, the mathematical theory makes no reference at all to the meaning of the messages whose transmission is the topic of inquiry, whereas an informational theory of meaning will have to consider the mathematical properties of information.[4] On the other hand, however, the mathematical theory of communication bases its concept of information on the notion of communicators of linguistic meanings, whereas informational theorists of meaning, being primarily interested in the origins of meaning, add qualifications to the role of agents as purposeful communicators. On these latter theories, information is, first and foremost, *natural* information, to be encountered and picked up by its receivers in their environments. It serves to first ground all sorts of intentional relations and hence cannot presuppose their presence. Purposeful transmission of signals comes later, historically and systematically.

The content of natural information, and some discontent

Retaining the Shannon-Weaverean presupposition that the presence and the activities of purposeful communicators of meaning are the basis of an analysis of information, Bar-Hillel and Carnap (1952), in their semantic theory of information, restricted the domain of their theory to sentences within a language. That domain includes not only natural languages but also arguably privileges propositions formulated in formal languages. As Bar-Hillel and Carnap argue in some formal detail, the information transmitted by a sentence simply *is* the content of that sentence. This point, however, cannot be readily transferred to cases where there is information but no sentences or other linguistic forms.

A rather different and influential take on the role of information in meaning was presented by Dretske (1981) – although it might be argued that the influence of Dretske's theory on philosophical debates is almost entirely composed of stimulating arguments *against* his view. Dretske's aim was to conceptually root intentional phenomena in general and the possibility of knowledge in particular in natural relations that he reconstructed as informational relations. These relations were couched in probabilistic terms that, although commencing from an interpretation of the mathematical theory of information, depart from it in important respects.

The obvious starting point for a reconstruction of Dretskean information is his near-infamous definition of informational content:

> Informational content: A signal r carries the information that s is F = The conditional probability of s's being F, given r (and k [standing for "knowledge"]), is 1 (but, given k alone, less than 1).
>
> (Dretske 1981, 65)

On this account, information is a certain relation between two world affairs s and r (things, events, properties or a combination thereof) in which r is a signal of the other affair s being or having property F only if and only when the signalled affair of s being F is the case whenever the signalling affair r occurs. Because there is a conditional probability involved in this definition, the relation in question must concern *types* of signals **r** that compose all the various *r*-tokens which obtain when *s* is *F*, so that a statistical correlation between *r*-tokens and *F*-conditions at *s* can be identified. Singular, non-repeatable co-occurrences of *r*, *s* and *F* cannot establish probabilities. This condition does not imply that informational relations can only hold for types of *F*-conditions at *s*, but it implies that there must be types of *r*-events that occur when some *s* assumes some property *F*. Thus, *F* may be a singular occurrence, or *s* may be an individual. The presence of informational content is strictly dependent on the presence of a correlation between *r*, viewed as members of types, and *s* being *F*.

At the same instance, Dretske's quest is for the content of a specific, individual signal, over which one cannot average, plainly and simply because it is an individual. It is either the case or it is not the case that *s* is *F* when some *r*-token is present. The conditional probabilities established for types of *r*-signals supposedly govern the individual instances. On Dretske's view, the correlations involved are of a strict kind, so that the conditional probability in his definition is $p = 1$. Whenever it turns out that *s*, instead of being *F*, is *G* or that *t* is *F* instead of *s*, any *r* that seemingly signals *s* as being *F* does not convey information, hence is not a signal, and accordingly has no content either.

To back up this claim, a second necessary condition for the presence of informational content is introduced, which is not contained in Dretske's definition but explicitly stated in the context (Dretske 1981, 73, 76f): *r* must be tied to the affair of *s* being *F* by some, at root nomological, regularity. Coincidentally parallel transformations of values at *s* and *r* will not count, even if the probability, in terms of frequency, is $p = 1$. Information is transmitted if and only if the conditions at the source affect the conditions on the receiving side in some regular way, that is in some way that stays the same over the full extension of instantiations of *r*.

If *r* may be caused by other conditions at the source than *s* being *F* (say *G*, *H*, etc.), or if *F*-conditions may hold for something other than *s* (so that *z* is *F*) but still cause *r*, or if *s* being *F* only intermittently causes *r*, no informational relation with the content that *s* is *F* obtains. Hence, causal relations are not sufficient for establishing informational relations, nor are they always necessary (Dretske 1981, 33–36). In turn, however, the relation between source *s* and signal *r* need not be a relation of condition *F* at *s causing r*. A causal relation between *s* and *r* may be indirect (Dretske 1981, 38). For example both *s* and *r* might have a presumably distal common causal antecedent that ensures covariance between these otherwise independent affairs.

If conditions at the source are to be reliably matched on the receiving side, stable "channel conditions" are required under which *r*-signals can be detected by *R* (Dretske 1981, 111–121). The presence and the nature of channel conditions will account for both the possibilities of information uptake by organisms and the kind and amount of informational noise that intervenes between source and receiver. It would not suffice for *r* to reliably covary with *s* being *F* if many or most signals could not reach a receiver *R*, for being lost in transmission and thus creating equivocation, or if the intervening noise added seeming *r*-signals that, in fact, did not originate at the source.

In an ideal situation, the channel neither subtracts information generated at the source nor adds extraneous information. In such an ideal situation, any transformation of conditions at the source would be matched by a transformation of the signals, and any distribution of probabilities for different outcomes at the source would be matched by an identical distribution of probabilities for different signals to obtain. This is, in fact, as much of an idealisation as a friction-less plane or point masses in the context of scientific modelling – whereas the central question for the mathematical theory of communication, as Dretske acknowledges, did not concern ideal situations but real-world problems caused by precisely those noisy channel conditions which one might just wish to idealise away: how much redundant information should be produced at the source so as to ensure that enough information will get through a channel affected by a specified amount of noise? That measure does not concern the reliability of information in terms of the probability with which *r* is generated in relation to *s* being *F*. Instead, it concerns the conditions for successful transmission of that information – which do not generate additional information about *s*. If channel conditions are variable, additional information on these conditions (rather than on what happens at *s*), ideally available through independent channels, will be useful to the receiver – that is *if* it is available.

It is the question of what happens in less-than-ideal situations that separates Dretske's theory from its main critiques. The critics claim either that the information-theoretical terminology of signals, channel conditions, noise and equivocation he mobilises in his account of informational content is misguiding in terms of bearing no tenable relation to the mathematical theory of communication from which it was borrowed (let me call this the "misnomer critique") or that Dretske's requirements for informational relations to obtain are too restrictive to meet real-world conditions (let me call this the "restrictiveness critique"). These two critiques are related and are best considered in relation to each other.

According to the "misnomer critique", as brought forward by authors such as Kenneth Sayre (1983), Dretske's focus on the content of a signal rather than the conditions for its successful transmission amounts to changing the topic, so his reference to that theory is tenuous or even purely rhetorical, and his liberal redefinitions of information-theoretical terms are prone to misguide the reader (Millikan expresses a similar view in personal communication). On slightly different grounds, but to similar effect, Brian Skyrms (2010, 34) argues that Dretske's semantic notion of information, in which informational content is conceived of in propositional form, relegates the insights of information theory to secondary importance.

The mathematical theory of communication confined itself mostly to measuring the signal strength and channel conditions required for successful transmission of some signal, whatever meaning it may have. Determining that meaning was considered an affair of senders and receivers of signals, not of objective relations between world affairs that do not *per se* presuppose the presence of a sender or receiver. The relations Dretske had in mind could be detected by a receiver, and they might be communicated by a sender, but in order to do so and convey information to them, the relations between source conditions and signals will have to be of a certain kind, independently of any sender or receiver.

The "restrictiveness critique", as brought forward by authors such as Ruth Millikan (2001; 2004) and Skyrms (2010) maintains, as Dretske later acknowledged, that setting admission criteria to the realm of informational relations as strict as Dretske originally did might leave us with an account of information that is quite removed from the ways in which organisms successfully perceive and act under less tightly bounded but more natural conditions. Reliance on relations between source and signal that are not nomologically governed but more probabilistic and local in kind is a common practice in nature. Some, and arguably most, conditions in nature will hold only within locally or temporally circumscribed domains, possibly with no method available to the organism for determining the boundaries of these domains, let alone the degree of reliability of some putatively informative signal.

If, on the one hand, one requires a strict, nomologically rooted covariance between the presence of *r* and *s* being *F* for there to be information at all, and if, on the other hand, it shall be acknowledged that natural organisms live under less-than-ideal, variable and often uncooperative conditions, the probabilities considered on either side, although related, have different reference points. Information as a probabilistic affair, in the manner introduced in IN-1 and discussed by the mathematical theorists, concerns only the question of how many binary decisions or "bits" are required to reduce a given number of antecedent possibilities to one determinate state at the source, and how much redundancy will have to be added at the source in order to compensate for noise-induced loss of that information during transmission, so as to avoid or minimise equivocation on the receiving side.

In contrast, the probabilities introduced by Dretske in his definition of informational content concern the question of whether a signal transmits information on a certain world affair, namely that *s* is indeed *F*. The question here is whether some affair *r* is related to what happens at *s* with sufficient reliability and regularity:

> The conditional probabilities used to compute noise, equivocation, and amount of transmitted information (and therefore the conditional probabilities defining the informational content of the signal) are all determined by the lawful relations that exist between source and signal.
>
> (Dretske 1981, 77)

That conditions at the source lawfully affect the signalling relation is the one core presupposition of the mathematical theory of communication – which is not further explained or elucidated in that theory. This presupposition is one key reason for Dretske to actually mobilise that theory, as he most clearly states in (Dretske 1983, 56): "For what this theory [= communication theory] tells us is that the amount of information at *r* about *s* is a function of the *degree of lawful (nomic) dependence* between conditions at these two points." Still, that dependence between conditions at *s* and *r* tells us something only about the measure of probabilities in a signalling relation.

The difference in focus here can be illustrated as follows: there may well be an *r*-token that signals "conditions at *s* unchanged". This would be an informationally rather bland affair on the mathematical theory, as the probability at t_0 of *s* being or becoming F at t_1 would be $p = 1$, so that no binary decisions are required and no information would be generated. However, to the extent that this signal is tied to that condition at the source unequivocally and lawfully, it is this kind of relation that is relevant to Dretske, in that only such a relation is able to ground knowledge. In turn, the transmission conditions investigated by the mathematical theory determine how strong a signal has to be, how much redundancy is required and hence what degree of reduction of possibilities at the source has to obtain in order for it to successfully reach its receiver once it relates to conditions at the source in an appropriate way. Both conditions have to be fulfilled in order to afford knowledge to the receiver of information. In fact, all the remarkably steep conditions Dretske imposes upon his notion of informational content are best read in the light of his ultimate purpose, which is a genuinely and rather traditionally epistemological one: what are the conditions of knowledge?

On Dretske's account of informational content, the probability in question is explicitly designed as an all-or-nothing affair: if p is not 1, it is bound to be 0 (Dretske 1981, 60). It is either true or false that s is F when some r-token that signals s being F is produced and transmitted, and if it is false, no information is transmitted. Only under these clear-cut conditions, Dretske continues to argue, can information afford knowledge. If information were allowed to be false or uncertain, and still be information proper, there would be no knowledge – notwithstanding the possibility of poor channel conditions or faulty mechanisms of information uptake that add uncertainty to its use. We could not even be properly mistaken about something in the first place if the underlying informational relations were already equivocal. Hence, the probabilities concerning informational content and those concerning information transmission uptake should be clearly distinguished – a distinction that is not always clear in Dretske.

To return to the example of frog vision that opened this chapter: a certain pattern on the frog's retina will elicit a certain behavioural response, namely the tongue darting out to catch and eat the object so projected. Under normal conditions in frog habitats, and under normal conditions of functioning for the frog's perceptual apparatus, the respective patterns are virtually always caused by insects, and they are so caused in accordance with fairly robust natural regularities. Given that insects provide nourishment to the frog, one would be entitled to say that the patterns on the frog's retina convey information on the presence of insects-as-nourishing-objects in his vicinity. However, if there is something dysfunctional about the frog's perceptual apparatus or if unusual external conditions intervene, the informational relation in question cannot be tracked. The perceptual mechanisms involved may well maintain their functions even if they fail to perform it. If an experimenter tosses lead pellets across the frog's visual field, any object causing a sufficiently similar pattern on the frog's retina would be met with the same response. The experimenter would be tampering with the channel conditions, adding seeming signals that do not have their proper source. Hence, the visual pattern would not convey information on the presence of insects anymore (Dretske 1981, 33–35).[5] If, however, the original informational relation were not firmly in place, the frog *could* not react appropriately, even under optimal channel conditions.

So here are the points that I see in Dretske's theory that are worthy of being 'resurrected': first, information is, above all, natural information, in terms of being present in the environment independently of any means of signal transmission or processing, and independently of senders and receivers. Second, and conversely, informational relations serve to ground content for their receivers. Third, information is relationally defined, in terms of lawful covariance between r and conditions at s, with no reified informational entities being postulated. The term "information" denotes a set of invariant relations in the environment that is foundational to the possibility of keeping track of objects and events throughout a multitude of conditions for a variety of differently constituted organisms. Hence, one key task will be to tell the variant from the invariant relations in this context. However, the question remains of how to account for the possibility that, under certain conditions, less-than-unequivocal relations (in which $p < 1$ for s being F, given r) might be of informational value to such a variety of differently constituted organisms, and that these conditions are the normal conditions under which perceiving and cognising organisms operate. Knowledge might not even be the gold standard on which to evaluate these conditions.

Alternative views of information

In contrast to Dretske's rigid approach to probabilities concerning informational content, Skyrms uses a more relaxed and expressly game-theoretic approach of "signalling games" between senders and receivers (Skyrms 2010, 35). On his account, the informational content of a signal shall be determined from the probability with which tokenings of the signal, in relation to some world affair, will influence the behaviour of the receiver (Skyrms 2010, 33–41). The direction and magnitude of that influence (the "vector") plainly *is* the informational content of the signal. The content vector determines the meaning of a signal within the context of a specific signalling game that applies to a given situation for the organism using and, in a subset of cases, also the organism producing the information in question. The direction of the vector might be objectively measurable, but it is a direction relative to the organism involved. His constitution, his abilities and his aims need to be taken into account. The information effects something in relation to him, and what it effects – in most cases, the adaptive response on individual or population levels – is the content of that information.

If a signal always and exclusively occurs in correlation with one of n possible states of affairs, its content is unequivocally determined within the context of the game, as is the response from the receiver. If the relation is less than unequivocal while still being identifiable as a vector of the aforementioned kind, the information transmitted is weaker, but it does not cease to be information. On this view, the probabilities for informational content come in frequency-based degrees. Moreover, the probabilities involved are now all treated on one level. If, on the repeated occurrence of some signal, there is an effect on the behaviour of an organism or a population in terms of an increased probability of some kind of behaviour to obtain in comparison to all available alternatives, and hence if the occurrence of some *s* being *F* to a certain degree covaries with a reduction of possibilities in the receiver's behaviours between t_0 and t_1, an informational relation is established.

The game-theoretic aspect of Skyrms's theory is not supposed to be read too literally: the signalling games involving senders and receivers of information are introduced in order to provide a formal model of patterns of modifications of probabilities that can be detected even in kinds of natural processes that could not possibly involve any consciously pursued strategies of players. For the senders and receivers, there is no presupposition of a "pre-existing mental language", and the analysis shall include cases "where no plausible account of mental life is available" (Skyrms 2010, 7, endorsing David Lewis's 1969 account of sender-receiver games).

It does not become entirely clear to what extent Skyrms introduces these notions as formal analogies that shall help reconstructing informational relations of all kinds, or whether they are to be read in a more material sense, where senders and receivers are involved in sending and receiving signals even in cases where they are not endowed with a mental life. At any rate, the probabilities involved are meant to be understood as frequency distributions of certain outcomes, not as degrees of belief on the sender's or receiver's side. What, however, *is* presupposed whenever real-world organisms are involved is some kind of adaptive dynamics that accounts for the transition from randomly variant towards directed and world-fitting behaviours. Variation and natural selection within populations are a necessary precondition of the evolution of signalling games. Successful behaviours and failures are accrued by real-world organisms, and provide the values on the basis of which a strategy is determined within the context of the game-theoretical model. Even though the players of signalling games need not be aware of certain strategy or of following that strategy, they will remain players in terms of their real-world behaviour and its consequences, as they are influenced by signals received and, sometimes, emitted by them.

Even more explicit reliance on the Lewisian notion of sender-receiver games is displayed in recent work by Peter Godfrey-Smith, who focuses on the co-evolution, broadly conceived, of sender and receiver behaviour – which may also comprise functional components of one organism. Instead of the uptake of natural information encountered in the environment, the paradigm applications of this model are, first, mechanisms of signalling within an organism, more specifically mechanisms of transmission of genetic information and of memory (Godfrey-Smith 2013b) and, second, rule-governed activities of signalling between organisms, in the context of animal communication (Godfrey-Smith 2013a). In the following chapters, I will make a case for environmental information for perception being the genuine paradigm of information, from chemo- or magnetotaxis onward, with active signalling within or between organisms building upon such natural information.

In a different critical take on Dretske's approach to the role of probability in information, Millikan (2001) observes that, if the probability required for informational content shall be $p = 1$, *r* would have to not only be a signal of *s* being *F* but also, at the same instance, transmit a mark of its reliability to its consumer, thereby confirming that it indeed is a signal of that very affair in a given context. If we briefly return to Dretske's discussion of channel conditions, however, we will see that the requirement looks slightly different: if the channel conditions are variable, the animal will have to use a different, independent channel that either confirms or disconfirms the original information or that transmits information on the conditions in the former channel. Meta-information of this kind is separate information and is to be assessed separately.

In either case, however, the consumer may not have additional means of independently verifying whether conditions hold under which a series of *r*-tokens or a single instantiation of *r* in his or her stream of perceptions transmits the information that *s* is *F*, nor does he or she necessarily have independent means of verifying that conditions in his or her habitat are normal. Little knowledge on these matters can be presumed on the side of the frog in our example. On a higher plane, human cognition faces the same problem. In any given perceptual situation, we cannot independently verify whether our perceptions are reliably connected with their distal objects – which Dretske does not consider problematic because such uncertainty does not affect the conditional probabilities of the informational relations as such (Dretske 1981, 56).

As an alternative to Dretske's view and the problem of the mark of reliability it seems to incur, Millikan (2001) suggests a decoupling of those probabilities which hold for informational correlations at the source and those which apply to the channel conditions, and thus to the noise and equivocation that will inevitably enter into the transmission of the signal and hence reduce the reliability of the information to the consumer. In prima facie analogy to Dretske, the probabilities involved hence are treated separately again, but in a fashion that differs in one important respect from Dretske's account – namely in what status accrues to probabilities $p < 1$ for the co-occurrence of *r*, *s* and *F*.

In order to accommodate for the possibility of probabilities to be lower than one in informational relations, Millikan goes on to distinguish between two types of natural information: the first, christened "information_L" (or "context-free information" in Millikan 2004, 35), is based, in Dretske's spirit, on correlations at the source that hold with a probability of 1 by virtue of natural necessities. However, Millikan continues to argue, there are not many correlations of this kind that would be within a normal consumer's reach. Instead, all organisms rely on a 'softer' kind of information most of the time, which she christened "information_C" ("local information" in Millikan 2004, 35). Information_C is based on local statistical frequencies that may well remain below a probability of 1. These statistical frequencies reflect a history of co-occurrences between signalling and signalled affairs and thus are determined on a different base from Dretske's in terms of underlying regularities. These regularities are not nomologically but historically and ecologically grounded. The statistical frequencies Millikan has in mind are hence also determined on a different sample of occurrences, namely local, spatio-temporally bounded ones, and hence subject to variation.

Actually, the statistical frequencies for the signal to actually occur in relation to the signalled affair might even be extremely low without undermining the informational relation as such. Millikan (2004, 71f) argues that, if there are enough successful instances of signalling a certain world affair to outweigh the effects of wasted effort or damaging mistakes, so that the ratio between the effects of the successful instances and those of the failures contributes to an explanation of the individual's and/or its species' persistence and reproduction, the successful instances in which *r* indeed signals *s* being *F* will suffice to determine the informational content of the signal.

In turn, the separation of the probabilities for the source and the channel conditions highlighted earlier is designed to specifically account for the possibility of failing instantiations that are due not to unreliable conditions at the source or dysfunctions in the receiver but to uncooperative environmental conditions that, varying over space and time in kind and intensity, may negatively affect the transmission. Under real-world conditions, less-than-perfect consumers of information will have to make do with abundant but incompletely trackable information in less-than-noise-free environments. Formal idealisations of the kind Dretske appears to endorse may contribute to missing this point.

Despite these differences, a common concern for Dretske, Skyrms and Millikan lies in providing an account of the possibility of meaningful yet false, misrepresenting signs. They diverge on the question whether there can be such a thing as false natural information to begin with. Where Dretske holds that, if it is false, it is not information, Millikan maintains that, if it can be false, it is not natural but already intentional information, while Skyrms settles for a kind of laissez-faire probabilism that allows for all degrees of reliability of information, the only threshold being the possibility of a breakdown of the hypothetical or real sender-receiver game. In fact, Millikan claims that the mark of distinction of the entire teleosemantic programme does not as much lie in the theories of meaning it proposes but in its specific set of theories of *mis*representation (Millikan 2004, Chapter 5). Depending on whether one counts Dretske and Skyrms as teleosemanticists, the variety of such theories within that camp seems to be fairly broad. As we will see in the next section, these differences concerning false information may ultimately betray a mostly implicit disagreement about a less obvious but more foundational set of philosophical presuppositions.

Natural information and the roots of intentionality

The fact that Dretske, when discussing informational content, treats probabilities at the source and probabilities concerning the channel conditions separately and sets the standards for the former improbably high is, with some likeliness, systematically grounded: it is the very detachment of knowledge-conveying information from meaning and its characteristic side effects of intensionality and error on which Dretske's approach builds. In his original definition of informational content, the consumer of information appears only as the unnamed bearer of the knowledge *k* in relation to which, under certain conditions, a signal *r* becomes the information that *s* is *F*. Under some, but not all, circumstances, *k* will become relevant to the informational relation, and only under some, but not all, circumstances, fully fledged propositional knowledge, or even any knowledge, is the product of the informational relation.[6] The addition to the definition of the parenthetical and – at least seemingly – optional condition of the presence of knowledge on the receiver's side, and the reference to a "suitably placed observer" who "could learn something about X" by consulting the information conferred by some state of affairs (Dretske 1981, 45), might distract the reader from the systematic point that he places at the centre of his inquiry (Dretske 1981, vii): that for natural information to exist, there is no need for interpretive processes that confer meaning upon a signalled world affair, so that there is no presupposition of a system capable of delivering such interpretations to be involved in informational processes either. Receivers of information might be of much more primitive nature than that of being interpreters or "consumers". A receiver of information may simply react to a signal in the most mechanical fashion or simply record it without doing anything else with it, yet without signals reliably transmitting information about world affairs, there would be no knowledge and no reference point for meaning. Hence, in order to keep the distinction clear, Dretske locates the paradigm of information below not only the level of linguistic forms but also the level of any interpretation in even the most elementary or rudimentary of semantic terms.

Interpretation of information, on this view, is a higher-level affair, where a signal not only triggers a reaction, but is recognised, consciously or otherwise, *as* a bearer of some information. Only on this level, intentionality and error will make their appearance. The world abounds with information, and the organisms' task lies in tapping into this resource, depending on their respective constitution and abilities. If an organism is unsuccessful in providing himself with reliable information on what is relevant to him, the blame will never lie with the presumed informational relations as such. If those relations are not reliable, they are not informational relations in the first place. Instead, the organism's own misfortunes, deficiencies or dysfunctions will be the first thing to be held responsible for any unhappy outcome that ensues.[7] In order to balance for this consumer-unfriendly bias, Dretske supplements his account of natural information with an elaborate theory of misrepresentations, as they will occur once we move from the realm of abstractly defined correlations towards the realm of biological *functions* of representing these correlations. Fulfilment of these functions may not only be thwarted by internally caused failures to perform them but also by abnormal conditions in the environment, Dretske (1986) later admitted – but this does little to undermine the foundational role of genuinely misrepresentation-free natural information.

What is at issue here is, on a proximate level, the difference between "informational semantics" as proposed by Dretske himself, Jerry Fodor, Dennis Stampe and other authors, and the "consumer semantics" brought forward by Millikan, David Papineau, Peter Carruthers, probably also Skyrms and others (see Dretske's own account of the distinction in Dretske 2009): although both approaches are of an externalist kind, the latter authors prefer to locate the meaning of linguistic signs or other signals 'downstream' from the source of information, namely in its effects on the receiver and his or her behaviours, whereas the former insist on grounding meaning in the original informational relations at the source.

If, however, we follow Daniel Dennett (1987, Chapter 8, "Evolution, Intentionality, and Error") in his account of the "Great Divide" in the contemporary philosophy of mind, the difference at issue here may ultimately be more fundamental than between systematically grounded preferences for variant perspectives and variant methods. Only to the extent that Millikan and other authors view information from the perspective of its consumer, that is the receiving and interpreting organism, and take this to be the relevant perspective for a theory of meaning, the problem of markers or meta-information on the reliability of the signal arises – which is to be solved by a functional-historical account of organisms dealing with signals that, with varying degrees of reliability, are related to world affairs. Whether by phylogenetic or by ontogenetic means, they acquire mechanisms for coping with those varying degrees of reliability and for tapping into other information channels that help to assess the affairs on the source side. Not so for Dretske and the authors Dennett subsumes on the other side of the Great Divide: the firmness and consumer-independence, if I may call it so, of the informational relations that Dretske insists on shall account for a solid grounding of whatever further operations in the organism's nervous system will obtain in terms of representing world affairs. There is an indefinite amount of natural information, and because that information is an entirely objective affair, all the information there is could be, in principle at least, assessed with determinacy. Only under this condition of a firm, nomologically governed, rooting in natural informational relations, symbols can become meaningful, and be endowed with what John Searle (1983) calls "intrinsic intentionality", also known as "original intentionality" – a notion that, on Dennett's account, Dretske endorses (Dennett 1987, 288f) and that, if tenable, would spell bad news for computational models and simulations of meaning-conveying processes.

Despite its rooting in natural relations, this latter view implies the assumption of a discontinuity between intrinsic intentionality and the definition and characterisation of normal biological traits. Organic traits allow for the existence of intermediate and proto-forms and thus for some degree of ambiguity in definition and function. The debates within and about evolutionary theory will never ultimately resolve, nor can they possibly evade, questions of delimitation between biological characteristics, such as, for example: at what stage in the course of evolution does a (dinosaur's) arm become a (bird's) wing? Is the demarcation between species determined by variance of traits, too, or by reproductive isolation only? When precisely does a speciation event occur? Is evolution always a continuous process? Even the Cambrian Explosion and Punctuated Equilibria, viewed in relation to the time-frame of evolution, remain gradual processes, albeit with a steeper gradient.[8] In these processes, there are no unequivocal, predetermined thresholds to tell us when something really has become something else, and when a fast-paced process really amounts to a rupture in the course of evolution. In contrast, the phenomenon of intrinsic intentionality, as conceived of by of Searle and like-minded authors, makes its appearance as an either-or affair. It will be difficult to imagine it as being partially realised or as an intentionality-in-the-making. Nor is it unequivocally admitted to being a product of processes of adaptation by natural selection in the first place. An explicit plea for such a discontinuity that is based on an explicit refusal to accept arguments for naturalising intentionality in Darwinian terms, and that, unsurprisingly perhaps, explicitly recurs to heterodox, non-adaptationist strands in contemporary evolutionary reasoning, is provided by Fodor (1996) in a polemical review of Dennett's *Darwin's Dangerous Idea* (1995). In contrast to the "adaptationist programme" (as understood by Fodor but not Dennett, who vigorously defends a different interpretation),[9] natural selection is not a teleological process in terms of biological traits being selected *for*, that is selected with respect to pre-existing criteria of fitness that may not (yet) have a counterpart in real-world conditions. Even if there were such a teleological element to natural selection, Fodor continues, that element would not suffice to ground the intentional character of thought – whereas the teleosemanticist is supposedly committed to believing that the intentionality of thought is derived, in terms of necessary and sufficient conditions, from some presumed intentionality of natural selection.

Whatever teleology there might be to natural selection, Fodor claims, it cannot, as a matter of biological fact, consist in being directed towards anything that is not actual, whereas the intentionality of the mind is characterised by the thoughts being directed towards real, ideal, prospective, imaginary or other states of affairs that, on Franz Brentano's famous definition, specifically and sometimes only exist *in* those thoughts (and hence *in*-exist, 1874, 115f). Fodor concludes that there must be something particular about the intentionality of thoughts that is not covered by a teleological nature of natural selection. This does nothing to speak against semantic naturalism, since Fodor, like Dretske, not only acknowledges but endorses the existence of natural representational content. The important qualification is that such content does not exist qua being selected for, but qua determinate natural correlations of the informational kind. Nor is Fodor's objection supposed to include the claim that intentional phenomena are not a product of evolutionary processes, including natural selection. However, he insists that it is not some intentional character of natural selection that makes them intentional.

Dennett's reply to this criticism in Dennett (1996) is every bit as instructive as Fodor's argument itself: the assumption that there is an element of intentionality to natural selection is a false allegation, so Fodor's critique appears to be based on an utter misreading of his and any serious evolutionist's work. Dennett insists that his point was precisely to demonstrate that intentionality and representation can be derived from processes that do not, and could not, be endowed with these same characteristics. Not only can teleological structures emerge without a conscious agent at the wheel, they can also emerge from processes that have no trace of goal-directedness to them. Goal-directedness begins with biological functions, not with natural selection, but natural selection is the causal antecedent of all biological functions, understood as "selected effects". It also is the historical antecedent of each individual biological function. The natural information an organism encounters in his environment becomes relevant to the organism only in the light of, and in relation to, these functions.

On these grounds, Dennett, like Millikan, goes on to claim that there are ever so slight degrees between the non-intentional and the intentional, in terms of the evolution of cognitive abilities, that it would be arbitrary to draw a line and decide where in the history of nature intentionality *really* begins. As all organic functions (which is not to say: all organic traits) are derived from processes of variation and natural selection, cognitive abilities, to the extent they do have a function, are derived from such processes, and in this circumscribed sense their endowment of intentionality – but not their concrete content – is derived from these processes, too. Hence, Dennett concludes:

> You have to give up original intentionality and see that all the late, robust, representation-wielding varieties of intentionality, both the words on the shopping list and the mental images in your head, are artefacts, and hence have derived intentionality.
>
> (Dennett 1996, 268)

It is the same idea that underlies any attempt at creating artificial systems that produce functional analogues to adaptive organic traits, including cognitive ones. Whether deliberately or not, Dretske's insistence, like Fodor's, on hard, nomologically grounded information amounts to a denial of this family of ideas. I will return to providing my own account of the probabilistic and context-bound nature of natural information (but I will not say much more about intentionality) in Chapters 4 and 6, after inquiring a bit more deeply into the role of natural information in perception, as it is controversially discussed in the literature. The key question in the following chapter will be how a reconciliation between a strict, Dretskean interpretation of informational relations with local, probabilistic conditions of information uptake in perception can be accomplished, and why it should be accomplished.

Notes

1 This is the basic concept of cybernetic machines as proposed and built by W. Ross Ashby (1960). His paradigmatic machine, called the "homeostat", was explicitly designed as a material model of the elementary mechanisms of adaptive behaviour in organisms.

2 A markedly different turn has been taken by John Maynard Smith (2000), who proposes a notion of information in the biological sciences that takes the original mathematical theory of communication to be rather directly and literally applicable to the relation between genes and ontogenetic development.

3 This semi-formalised description of information is based on Weaver (1949, 12) and Dretske (1981, 12f).

4 There is a fine distinction that already shows up in Shannon and Weaver (1949, 8, 31) and marks a divergence in opinion between the two authors: Shannon assumes a complete irrelevance of meaning to the mathematical theory of information, and seems to imply that, conversely, the mathematical theory of information is equally irrelevant to any account of meaning. In contrast, Weaver maintains that there is an asymmetric systematic relation between the two, obtaining from mathematical theory to theories of meaning, most explicitly in Weaver (1949); also see Dretske (1981, 41 and endnotes 2 and 3 to that passage).

5 This example is also discussed in similar contexts by Dennett (1987, Chapter 8), in an explicit critique of Dretske's and other authors' position on the phenomenon of misrepresentation, and in Millikan (1986, 71f) in an analysis of the conditions for an organism

of exercising his natural, evolutionary acquired proper functions; see further discussion on pp. 112–113.

6 With a somewhat different aim in mind, Millikan (2001, 106) suggests that we could drop the *k*-condition without any loss if we integrate *r* into a more complete signal that includes information about *s* previously transmitted to the receiver.

7 One socio-political connotation of this interpretation of Dretske's view will not escape the attentive reader: it reads like the semantic counterpart to a roughshod conservatism which I would paraphrase as follows: "If you make it in life, it's because you're quite smart, so you deserve it. If you don't make it, it's your own fault, so please don't blame unfavourable circumstances or, even worse, sinister forces of falsehood conspiring against the honest workingman's fortunes." (I am, however, convinced that Dretske never intended such a parallel.) Compare this to Millikan's emphasis on the importance of the environment and her emphasis on the possibility of its uncooperativeness – and the generally very favourable reception of her theories, albeit mostly on different grounds, among animal rights activists.

8 The theory of Punctuated Equilibria was introduced by Eldredge and Gould (1972) and received thorough criticisms from Dawkins (1983, 412–418) and Dennett (1995, Chapter 10.3). Both authors target the issue of choosing the appropriate time-frame. The clause "viewed in relation to the time-frame of evolution" is relevant in this context because, in a trivial sense, everything, even the ignition of a nuclear bomb or the Big Bang, is a gradual process – if we adjust the time-frame accordingly. The most elementary unit of evolutionary processes are generations of organisms. Definitionally, every modification that takes $n > 1$ generations to obtain and does not suddenly appear from one generation to the next may count as a gradual process in evolutionary terms. Factually, however, even for rather rapid or fairly minor developments, *n* will normally be much larger than 1. The Punctuated Equilibria debate concerned the distinction among, first, a notion of continuous evolution, second, alternating paces with periods of relative stasis punctuated by processes of relatively rapid change and, third, genuine evolutionary saltation, that is macro-mutation. These are different processes, occurring on different time scales, that should be, but not always are, viewed apart from each other.

9 The term "adaptationist programme" was coined, with critical intent, by Gould and Lewontin (1979). Fodor's critique builds on this seminal essay and adds some more radical tenets to it (which he continues to do in Fodor and Piattelli-Palmarini 2010). For the classical sources of Neo-Darwinian adaptationism adhered to by Dennett, see Dawkins (1983), Pittendrigh (1958), and Williams (1966); for a qualified and nuanced defence of adaptationism as a research programme against Gould and Lewontin (1979), see Sober (1998).

3 Varieties of perception

One implication of Fred Dretske's nomological, objectivist view of information is that his aforementioned "suitably placed observer" is not necessarily identical with the actual receiver or consumer of the information in question. He might be the observer of an informational situation in which a different receiver is involved. The receiver in that situation may or may not learn something about X, and he, she or it may or may not have access to information about the reliability of the signals received. Presumably, the frog belongs to the less privileged group. The human observer is taken to be in a better position, for being able to assess informational situations in a way that other organisms are not. This epistemic privilege seems to include, again in principle at least, the informational situations he, she or it is involved in – even if no ultimate verification of the relations of our perceptions to their distal objects is in reach. This epistemic privilege seems to be assumed without further justification, and it places human agents outside the environmental constraints that affect other organisms.

A second implication of Dretske's informational semantics is that, despite all his reference to the mathematical theory of information, his concept of information is at variance with notions of information in mathematics or computer science in at least one important respect: it may work as a remedy against the symbol grounding problem that keeps haunting AI, but provides no account of how natural information is processed. Conversely, AI, along with the mathematical theory of communication, often confines itself to the tasks of information processing, leaving the grounding of the symbols so processed to take care of itself, or assuming it to be someone else's business anyway. This observation does not rule out the possibility of a division of labour between computer scientists and AI researchers on the one side and approaches to cognition based on theories of natural information on the other. However, the very paradigm of a psychology of perception that is based on an account of environmentally rooted natural information, a paradigm endorsed by Dretske (1981, Chapter 6), is expressly sceptical of AI, in assuming that no processes of a computational kind are involved in perception to begin with: James Jerome Gibson's ecological theory of visual perception (1979). Although other proponents of theories of natural information are not as resolutely disinclined towards the computational realm, Gibson's stance may count as symptomatic of a systematic incompatibility.

These two implications of Dretske's view of information stand in an interesting but not immediately obvious relation to each other. Besides further inquiring into the nature of natural information in the context of perception and aligning the Dretskean with the Gibsonian concept of information, the meta-goal of what follows in this chapter is to show how the two implications of Dretske's view are connected – although probably not in terms of being mutually supporting constituents of his theory. It is the computationalist view of perception that is ultimately committed (or condemned?) to relying on an objectivist standpoint in which relations between a perceptual state and its object can be unambiguously determined – while they cannot be warranted by the computational models proper. Conversely, everything in the view from natural information, if and when pursued consistently, speaks for informational relations that are objective in kind but context-bound all the way down, with no epistemic privilege to be assumed for the human being perceiving and acting within an environment. In this chapter, I will discuss three theories of perception that are markedly at variance with each other in these respects: David Marr's computational view of visual perception, Gibson's ecological approach to visual perception as the former's main antagonist and the main focus of attention in this chapter, and the Empirical Strategy of perception as a contemporary theory of intermediate status, which endorses the notion of context-boundedness of perception while relying on computational models. The issues of illusion and misperception discussed in that latter section will be particularly instructive to elucidating the role of natural information and its environmental context in perception.

Perception as information processing: the computational view

One peculiarity of all accounts of natural information discussed so far is that they do not include any notion of data or information processing. Dretske, Brian Skyrms, Ruth Millikan, like Gibson, all use concepts of *natural* information that either rest on the expectation that all kinds of information are deducible from a basic theory of natural information (Dretske, Skyrms) or do not include any notion of likeness between natural information as used in perception and the information-processing kind of information (Gibson, Millikan) – as Anthony Chemero (2003b) has usefully highlighted.

Notions of data and information processing are characteristic of the non-semantic mathematical theory of communication. They are also used in other, non-naturalistic, semantic accounts of information, such as in "General Definition of Information" or GDI that Luciano Floridi (2011) refers to as an "operational standard" in fields such as data mining or information management. Floridi's GDI characterises semantic information as *well-formed and meaningful data* – a definition *prima facie* both more vague and more specific in character than Dretske's. It is more vague in having to be supplemented with further definitions capable to cash out the requirements of being well-formed and meaningful. It is more specific in focusing, by definition, on the kind of data that can be processed by computers. Although not formally restricted to data of this kind, a fairly material notion of signals that are transmitted from sender to receiver through information channels, as in computing and telecommunications, serves as the paradigm of semantic information. In fact, Floridi identifies the source disciplines of his GDI to lie in those fields which treat data and information as reified entities – almost literally: information is composed of bits and pieces (or bits and more bits for that matter) that are processed by an appropriate assembly of hardware and software so as to produce a certain output. No such reified entities can be found in theories of natural information, where information is a probabilistically described (but, in Dretske's case, nomologically governed) relation between world affairs that does not add anything, ontologically, to a natural environment and that assumes its status as information long before and perhaps without ever becoming processable by some appropriate, organic or other, machinery.

With respect to the use of information for perception by organisms, the processing view of information presumes that, in order for an organism to produce a structured response to some external stimulus, that stimulus, which taken by itself is insufficient to structure that response, has first to be encoded in perception and then modified through a series of processing stages, which are likely to include the addition of other information, coming from different sources and through different channels, so as to arrive at a fully image-like or propositionally structured representation that finally serves to inform the organisms's response. The information relevant to the organism is constructed in the process of perception. Marr was the advocate of such an explicitly computation- and AI-based view of visual perception and is aptly considered the founder of computational neuroscience. He not only was a leading opponent of the Gibsonian view in the field of the psychology of perception but also closely worked with leading figures in classical AI such as Seymour Papert and Marvin Minsky. In Marr (2010), which was first published in 1982, he conceived of vision as the process of constructing, on different levels of perceptual processing, descriptions, via internal representations, of the information derived from an input image. "Representation" and "description" are to be understood as technical terms in this context:

> A *representation* is a formal system for making explicit certain entities or types of information, together with a specification of how the system does this. And I shall call the result of using a representation to describe a given entity a *description* of the entity in that representation.
>
> (Marr 2010, 20, emphasis in original)

In Marr's own example, the Arabic or binary numeral system would be the representation that provides the elements and the rules by which to generate any string of elements (33, 42, 999, etc.) as the descriptions of the individual numbers (rather than the numbers themselves). Hence, a representation is not a mental image or a sentence that is supposed to refer to some world affair, and that would be generated in the perceptual process, but a formal method of generating symbolic descriptions of that world affair or the intermediate staged of the perceptual process. The perceptual process generated by representations, so understood, and resulting in descriptions, so understood, is divided into stages (summarised in Marr 2010, 36–38): from the representation of the two-dimensional retinal image, a "primal sketch" is generated as a symbolic description of properties of the input image in terms of "intensity changes" (edges, boundaries, virtual lines, etc.). On these grounds, surfaces, textures and their orientation are described and apparent motion is constructed. Only from there, a three-dimensional representation, now understood in a non-technical sense, is generated that describes shapes and spatial arrangements and that is centred on objects. Only at this last stage, conscious perception is accomplished. In this process, the physical characteristics of a "scene" are "recovered", that is inferred, from the image (Marr 2010, 330f). Those physical characteristics work as objective constraints upon perception – which remains otherwise underdetermined by what is present in the perceiving organism's surroundings.

Perception remains underdetermined in Marr's theory inasmuch as the retinal image from which the perceptual process commences is conceived of as a mere pattern that does not convey information, and certainly no informational content of the kind defined by Dretske, about the world. The pattern is made up of the distribution of intensity values and their transformations across the retinal image. These intensity changes are caused by the geometry and the reflectance of the visible surfaces, the illumination of the scene and the viewpoint (Marr 2010, 41). The distribution and changes of intensity values thus caused are all the perceptual system has for processing. Detection of intensity changes is accomplished by algorithmic operators that transform the input so as to produce values for mathematical functions with peaks, troughs and zero crossings as their output (Marr 2010, 54–73).

Quite self-evidently on this account, the same pattern and the same distribution of values for a set of mathematical functions may be caused by a variety of different things, so that the organism has to make inferences as to how the world stands from the analysis of these patterns in further stages of visual processing. This is the problem of the ambiguity of the retinal image or of inverse projection, as classically stated in George Berkeley's *Essay Towards a New Theory of Vision* (1709). Berkeley's initial observation was that distances cannot be directly perceived, and that space as such cannot be seen, and that similar conditions apply to other perceptual qualities, such as the magnitude of objects. Identical retinal images can be caused by various objects, under various conditions, in various constellations. Where the association of experiences was the main disambiguating factor in Berkeley, inference of spatial properties from the retinal image serves that purpose in Marr (as in the "geometrical" theories of visual perception by Descartes and Malebranche to which Berkeley's was intended as an alternative). If, however, such inference is insufficient for disambiguation, and thus if the frog has no means of telling apart the intensity changes for insects and lead pellets moving across his visual field, there is not much that Marr's theory could do about this unfortunately ambiguous state of affairs. Issues of this kind would have to be relayed to ecological or other biological theories that, very much independently from a theory of perception, account for how frog's environments are shaped and how changes in environmental conditions will affect the frog's survival and reproductive success.

Visual processing is one contributing factor to an organism's successful behaviour, but, taken by itself, it only allows for inferences as to how the world stands, which may or may not be borne out by how the world actually stands. These inferences are consolidated *for the perceiving organism* by the repeated success of his behaviours towards what is perceived over the course of repeated perceptual instances. They are consolidated *for the external observer* by matching different perceptual instances against each other and against the physical knowledge at hand. To some extent, this picture resembles W.V.O. Quine's image of "how the human subject of our study posits bodies and projects his physics from his data" on the one hand (1969a, 83), and of "total science" as "a field of force whose boundary conditions are experience" on the other (1961, 42).

In order to get this "cognitivist" view of visual perception as information processing to work as proposed by Marr, and in order to counter the underdetermination problem, information has to be reified, so as to keep a mark of identity for what is being processed throughout all stages involved while, at the same time, keeping it attached to the original stimulus and the outgoing response. Marr does not present an explicit concept of information, but only on a reading of information as entities that can be subject to a well-defined sequence of formal operations, traceable from input to output, the complex array of symbols and their transformations involved in perception will be properly grounded. And only to the extent that there *is* a parallel between representational and computational processes, in terms both of the nature of information and of the methods of processing involved, and in the same sense as computation *is* cognition to Pylyshyn (1980), the perceptual processing in Marr's theory has a claim for an empirical grounding (see the schematic representation of the parallel in Marr 2010, 332).

Where information is reified in Marr's account, its receivers, in certain respects, are not. Information processing, Marr maintains, must be understood on three clearly distinct levels: first, the goal of the computational process has to be established; second, the algorithms and the representations for input and output need to be determined; and, third, the physical realisation of representation and algorithms has to be identified. Although they will provide an understanding of some information-processing task only in conjunction, these levels are relatively independent of each other (Marr 2010, 24–27). Accordingly, and in alignment with the computational paradigm, the formal, algorithmic structure of the subject matter can be analysed in abstraction from its concrete physical realisation, and may be realised in a variety of physical arrangements. Hence, one could not only provide machine models and computer simulations of perceptual processes but also construct a "general purpose vision machine" (Marr 2010, 331). The concrete physical realisation of a perceptual system, being part of an organism interacting with his environment, is relevant to be sure, but remains underdetermined by the computational theory proper, and is described in abstract formal terms. Together with the presumed parallelism between perceptual and computational processes, it is this assumption that bears witness to the alignment of Marr's theory with AI and placed it at the origins of computational neuroscience.

Information specifies affordances: the ecological view

> Animals and humans communicate with cries, gestures, speech, pictures, writing, and television, but we cannot hope to understand perception in terms of these channels; it is quite the other way round.
>
> (Gibson 1979, 242)

If Marr concludes from the observation that "vision is the process of discovering from images what is present in the world" that vision is, "first and foremost, an information-processing task" (Marr 2010, 3), and hence a task that involves various levels of symbol manipulation and inference in the very process of discovering what is present in the world, we find very different conclusions being drawn from the same basic observation, and from one seemingly shared premise, in Gibson (1979): that an inquiry into visual perception will have to concern the uptake of information and the channels through which it arrives at the visual system. However, where Marr focuses on the complexities of processing that information as the key to his analysis, Gibson quite flatly denies that perception is a matter of information *processing* in the first place. Instead, he maintains, one will understand vision only if and only when starting from the relation of the perceiving organism to his environment and the information provided by that environment. Informational relations are ubiquitous and objective, and can be directly acted upon without stages of processing and inference intervening: "Information as here conceived is not transmitted or conveyed, does not consist of signals and messages, and does not entail a sender and a receiver" (Gibson 1979, 57).

Even though Gibson provides only a rudimentary positive definition of his concept of information when maintaining that it "refers to specification of the observer's environment, not to specification of the observer's receptors or sense organs" (Gibson 1979, 242), it is obvious enough that no concept of information could be further removed from the mathematical theory of information, and from its presumption that intentional agents are present at the source and receiving sides of information, and that they are in charge of the meaning of whatever is communicated.[1] If information is not symbolic in the first place, and if it is not processed in perception (whatever else is done with it in the process of perception), and if it is an objective affair that is present in the environment even in absence of sender or receiver, it will not be affected by issues of its proper grounding. Natural information is grounded *as such*, in the very environments as they are inhabited by organisms, and it can thus become the root of all perception and, by implication, all cognition and empirical knowledge.

It is unsurprising that Gibson's account of perception, unlike Marr's, did not exert much influence on cognitive science, let alone computational neuroscience, while having laid the foundation to an entire school of ecological, non-computational psychology of perception. Gibson's approach has also become a source of inspiration for naturalistic philosophers such as Dretske and Millikan, and it has exerted a strong, though mediated, influence on many domains of designing artefacts, from architecture to computer interfaces.[2] This divergence of influences did not come by coincidence, as Marr had while Gibson had not much to say about processes on the neuronal level of perception, whereas Gibson had while Marr had not much to say about the guidance of action by perception in an environment. The illustrations to each author's main work will provide a hint of the fundamental difference at issue here: the figures in Marr (2010) are, for the most part, patterns that are supposed to represent retinal images and the stages of their processing, circuit diagrams of the computational architecture of perception and function graphs of the algorithms that govern visual processing, whereas Gibson (1979) provides the reader with a number of semi-abstract, schematic visualisations of how ambient energies impinge on organisms placed in, and perceptually directed towards, an environment with certain properties.

Perception, according to Gibson's view, does not amount to an image-like representation of an outer physical world. One should not take experimental settings as the paradigm of perception, where the subject's visual apparatus is exposed to momentary stimuli detached from environmental settings ("snapshot" or "aperture vision", Gibson 1979, 1). Nor do pictures or other mediated representations of world affairs provide a suitable paradigm of perception, as the information they provide is confined to a few aspects of their subject matter, allowing the viewer to capture only a limited subset of the information available in the environment (Gibson 1979, Chapter 15). When moving in relation to a picture, one will discover the difference to a real scene with ease. It is the dynamics of spatial and somatic relations between perceiving organism and object that has to be systematically accounted for. This is a distinction that Marr could not have made in a principled way. An identical retinal image caused by a natural scene and a pictorial representation would be treated in identical fashion by the perceptual system. Only the context of that representation could account for the difference, but that context is not part of Marr's inquiry.

The idea that visual perception begins with the projection of an image onto the retina, Gibson holds, will be misguiding to begin with. Animals with compound eyes, that is animals whose eyes neither have a lens nor a retina but are composed of an array of closely packed light-sensitive tubes, can produce reasonably accurate visually guided behaviour without even the possibility of deriving their visual perceptions from retinal images, or equivalents thereof (Gibson 1979, 61f). The concept of a retinal image misguides us into believing, first, that we not only process such images but also actually see them, or that some instance in our brain could see them, and make inferences – which Gibson derides as the "'little man in the brain' theory of the retinal image" (Gibson 1979, 60). We might stop and reflect upon our perceptions by wondering what that red blot over here may be, or what shade of red it displays. Arguably, most philosophers think of perception in this fashion, but part of Gibson's mission is to demonstrate that this is not how perception *works*. Second, and more subtly, the notion of the retinal image suggests that we perceive stimuli, and that perception is a response to those stimuli in which we derive information from them. Information would come into play, or even would be generated, only here, in the stages of perceptual processing. However, there are perfectly conceivable situations in which there is an abundance of stimuli that does not convey any information, such as in a brightly lit room filled with dense fog (Gibson 1979, 52–55; see the discussion in Chemero 2003b) – or, to use a more commonplace example, occurrences of so-called whiteout conditions, which can be dangerous to pilots, motorists or mountaineers precisely for the combination of a strong stimulus with the utter lack of visual information. Stimulus and information are in this sense detached, whereas stimulus and perception are not.

First and foremost, perception, on the Gibsonian view, is to be considered an activity that is intrinsically tied to other activities of an organism, and that depends on his general constitution and abilities – his physiology, his body scale, the behaviours he is capable and the resources he is in need of – on the one hand, and on his current position and movements within his environment on the other. Perception consists in the "pickup" of information from the "ambient energies" surrounding the organism. In relation to his position and movements, these energies form the "optic array" for perception (Gibson 1979, Chapter 5). To accomplish the task of information pickup, visual and non-visual information about the position, orientation and movement of the perceiving organism is included in the act of perception (Gibson 1979, 115–120).

In this fashion, information is actively retrieved from the environment, so as to detect patterns of persistence and change therein and track the "invariants" of some object. Invariants are to be understood in analogy to the mathematical meaning of the term (Gibson 1971, 30; 1973), as those properties of an object which remain unchanged when a set of rule-governed transformations is applied to it. For example the length ratios of a geometric figure remain unchanged when it is scaled up and down proportionally. These ratios, but not the absolute measure of the figure's elements, are the object's invariants. In the context of perception, the transformations will encompass all naturally occurring changes in the conditions of perception, and the invariants will be what remains unchanged, as viewed in relation to the transformations of these conditions. In Gibson (Gibson 1979, 45), we find a non-comprehensive list of candidate invariants, which comprises "alignment or straightness [. . .] as against bentness or curvature; perpendicularity or rectangularity; parallelity as against convergence; intersections; closures and symmetries". Citing Gibson's own example (1979, 13), a solid substance is rather persistent in shape, so shape is an invariant in the perception of all solid objects, but not in the perception of any less-than-solid object, for which density or volume are likely to count as invariants.

The acts of retrieving information from the environment and hence tracking the invariants of some object or event do not involve the "replication" or "copying" of that object or event in the ambient light, as though some replica of the object were picked up in perception (Gibson 1979, 102f). The tracking of invariants is much less concerned with detecting similarities between an image and an object than with guiding the perceiving organism's activities towards that object. That guidance has to be accomplished throughout a multitude of transformations of conditions within the environment.

We may find Gibson barking up a largely uninhabited tree of naïve metaphysics when he thus insinuates that other psychologists and philosophers of perception actually believe in a "copy" theory of perception, but his underlying point is quite remarkable: that perception does not involve representation. It neither builds on image-like representations, nor does it rely on generating an internal model that would represent the relevant variables that determine the appearance of an object, or on classifying objects or on otherwise abstracting from their concrete properties – all the things the computational approach to visual perception cannot do without.

Above all, the organism's acts of perception imply the direct uptake of information on what can be done with a perceived object. Throughout all instances of being perceived, objects of all kinds, including other organisms, to use Gibson's most used (and often misused) concept, provide *affordances* to the perceiving organism. An affordance is what some object offers to be done with it, in the particular way in which it is related to the perceiving organism at a given time, under a given set of conditions. Accordingly, an object's affordance is always described in terms of possible actions of, or interactions with, the perceiving organism, such as standing, sitting, climbing up, jumping over, falling off. For example a fruit may afford activities such as being eaten, burrowing, being picked and used as a missile or poisoning – or simply nothing at all. Affordances will vary among different animals for one and the same object or even the same organism on different occasions. The activities an object affords to an organism constitute its specific "values" or "meanings", which can be directly perceived by the organism (Gibson 1979, 127).

It would be unfortunate though to simply equate affordances with values and meanings. These are normatively highly charged terms that are likely to result in philosophical entanglements – which to avoid was one of the purposes behind the introduction of the concept of affordances (see Gibson 1966, 285). However, there are at least two related points at which that concept appears so vaguely circumscribed in Gibson's famous definition (1979, 127, 129) as to invite such entanglements: there, he says that "*affordances* of the environment are what it *offers* the animal" (emphasis in original), but he also says that an affordance is "neither an objective property nor a subjective property; or it is both if you like" and that it is "equally a fact of the environment and a fact of behaviour". So, first, are affordances *properties* of *objects*, or are they better defined in terms of relations and situations? On an ecological view, objects and their properties might, and should, not be considered in detachment from the environments in which they are encountered, so Gibson's reference to objects and their properties should be read in the light of this basic commitment. Second, are affordances entirely provided by the environment, or are they (partly) subjectively defined? If they are something that can be detected by an organism, they are supposed to be contained in the environment, but what is actually afforded in the act of perception can be determined only by reference to the constitution and abilities of the perceiver. Although the second apparent vagueness can be demonstrated to have a systematic purpose behind it, the first might be more difficult to parse.

Gibson's ambiguous characterisation of affordances as *properties* and as *facts* might seem a minor issue, but the implications are significant if facts are taken to be composed of things, properties and their relations. Context indicates that a reading of affordances as relationally defined facts rather than properties will be the most appropriate one. If affordances, their meanings and values were plainly and entirely properties of the environment and the objects therein, we would not only have to accept that such normative qualities were constituents of the environment, but also we would end up in a world cluttered with an indefinite array of such qualities that would have to be embodied or embedded in the objects and environments in which they are encountered, readily pre-packaged for all possible perceivers and detectable under all natural conditions of perception. These qualities might even contradict each other in certain cases, for certain perceivers, and still would have to be granted the same ontological status.

If one, disheartened by this prospect, does not want to settle for the opposite view that all normative qualities of some perceived object, and all variance therein, are located inside the organism proper and hence a subjective affair, an alternative is to locate values, their specificity to the perceiver and their variance between perceivers in the *relations* between organism and environment. Affordances, as Gibson insists, are always related to activities, and are preferably described as capacities or abilities of the object with respect to the perceiving organism's constitution and abilities. The chair has the capacity of getting me seated, and the cliff has the ability to make me fall off. Even if we render these facts as seeming properties of fall-off-ability or sit-on-ability, these can be defined only in relation to the animal – with rather different outcomes for ants and primates. These relations will be best captured by a more complex, fact-like description, such as "the chair is sit-on-able for humans" or "the cliff is fall-off-able for large mammals".

Hence, it is a relational interpretation of affordances that appears best to make sense of that other, purposeful vagueness in Gibson's characterisation of affordances as being objective and subjective at once. This relational, bi-directional interpretation can be further substantiated by an analysis of the nature of the information involved in affordances as being equally relational and bi-directional. This is what Gibson himself suggests when says that both affordances *and* information point "two ways, to the environment and to the observer" (1979, 141).

On this reconstruction of Gibson's notion of affordances, it is both realist and anti-dualist in philosophical spirit. This reconstruction should carry over to an account of ecological information for perception. There is one seeming ambiguity in Gibson's underdefined account of information that should be taken to be systematic.[3] On the one hand, "information [is present] in ambient light to specify affordances" (Gibson 1979, 143), and as such is present and specific even if and when an affordance is not perceived. On the other hand, information for perception is always composed both of relations *within* the environment and of the perceiving organism's relations *to* his environment. Hence, an important part of information for perception is dependent on the perceiving organism, and all information for perception is relational in kind. Within ecological psychology, there has been a variety of diverging interpretations of the relation between Gibsonian affordances and information. Affordances are viewed as properties by some (Heft 2001; Reed 1988; Turvey 1992), as relationally defined by others (Chemero 2003a; 2003b); information is considered properly external to the organism by few (Reed 1988), and as organism involving by most others (Chemero 2003b; Costall 2004; Heft 2001; Turvey 1992; 2013; van Dijk et al. 2015).

In particular, Michael Turvey refers to affordances as dispositional properties of objects that are actualised under a concrete set of conditions, whereas Chemero (2003a) develops a relational theory of affordances that is complemented by an equally relational account of information for perception (Chemero 2003b). Whereas relations exist between entities with certain properties, this does not entail that these relations are properties themselves. For example John is taller than Sally. Quite obviously, "being taller than" is not an absolute measure, nor is there a complex property of taller-than-ness, let alone taller-than-Sally-ness possessed by John (Chemero 2003a, 187). If we conceive of information as informational *relations* from the start, we do not incur the ontological challenge of justifying such complex properties. If perception of an affordance is based on relations between the ambient energies and structures in the environment, including the perceiving organism himself (Chemero 2003b, 580f), an affordance is, in Chemero's wording, a relational "feature" of entire "situations" of an organism moving in and through a certain environment. The information picked up in perception does include relations between organism and environment. It specifies the environment *and* the self, and hence can be both an objective commodity *and* relationally defined.

Under this interpretation at least, Gibsonian information for perception assumes an appearance quite similar to Dretske's notion of information. Although he does not clearly define it as such (and although Fodor and Pylyshyn 1981, 166–167 claim otherwise), Gibson appears to employ a view of information that is similarly realist and relational when he discusses the perceiving organism's tracking of the invariants of some object in the environment and the activity of extracting information from the ambient optic array. If that information is supposed to remain identical throughout the various instances and situations of perception, it has to be regular in a fashion that is independent of subjective conditions – even though the perceiving organism's position, constitution and abilities will enter into the set of perceivable relations. As these relations remain stable and unequivocal in an otherwise changeable environment, he can relate to them in various ways over time, under variant conditions. Although organism-related variables enter into a given set of informational relation that he uses, the functional status of his sensory organs or his degree of attentiveness in a given perceptual situation will not. Only if the informational relations and their stability can be taken for granted will there be affordances that are organism specific.

An object affords what it affords because some of its properties remain stable and reliably detectable in the ambient optic array for the perceiving organism. They are reliably detectable precisely if there are, first, conditions for perception which are such that any transformation of the object or its position has strict and unequivocal correlates in its placement in the ambient optic array. Second, the perceiving organism is able to match his position, movements and aims against the relevant environmental invariants over time, thereby using proprioceptual information. Both sets of relations in conjunction are the informational relations that specify, in Gibson's phrasing, the environment for the organism. Third, the organism must be in a condition suitable to actually picking up that information. Precisely if and when both the environment- and the organism-bound conditions are fulfilled, there will be a direct perception of affordances: the perceiving organism is enabled to immediately rely on the relation between himself and what remains invariant, what varies regularly and what varies arbitrarily in an object over the course of his perceptual activities.

For thus including organism-related variables, information in the ecological sense is "intrinsic" to a perceptual situation, as distinguished from "extrinsic" information that allows for absolute, perceiver-independent measurements of physical and physiological variables by an external observer. Intrinsic, affordance-related information still allows for measurement, although on a different basis (see Gibson 1979, 128). The paradigm of measurement-oriented approaches in post-Gibsonian ecological psychology are the measurements of stair-climbing affordances for persons of different leg lengths in William J. Warren's study (1984, where the intrinsic/extrinsic distinction is introduced; see also Boumans 2013). Although there is considerable variance in the relation between riser heights and leg length for short and tall persons, the results of the experiments in this study suggest that there are ratios for stair-climbing affordances that remain constant across the study population. These ratios can be expressed in an intrinsic or body-scaled metric that matches the dimensions of the subjects' bodies and the dimensions of the object in question – in this case, the riser height of the stairs. Such ratios can be found both for perceptual category boundaries – between stairs that are perceived as still being climbable and stairs that are already perceived as being unclimbable – and for optimal heights in terms of effort to be invested into climbing. The ratios are such that short people will perceive the same affordance for relatively low risers that will be perceived by tall people for proportionally higher risers. The ratios, however, do not express subjective factors, in that they remain constant across the population, with identical absolute measures for people of the same size, and in that perceived optimal heights quite closely match the actual energetic optima in physiologically based trials. Hence, Warren concludes, environmental objects are perceived by an organism in relation to his action capabilities, where that relation is rather stable and unequivocal.

On this background, affordances can be relationally defined, in a two-step fashion that builds upon informational relations and the perceiving organisms' abilities to track, and act upon, these relations – which will include his own specific relations to his environment. This relational view allows for a fairly clear set of distinctions to be made with respect to situations in which an affordance, as it is perceived by an organism, and information for perception do not appear to match: there is no information if the relations between signalling and signalled world affairs are not unequivocal, even if and when an organism appears to track such a relation, so there is no affordance to be perceived either. Conversely, if there is a malfunction in an organism's perceptual system, information will still be present in the environment, as there will be nomologically governed relations between incoming light, refraction and reflection on surfaces and so forth, but these relations could not, or not properly, be detected by the perceiving organism for reasons internal to him. Hence, the third of the previous requirements, namely that of being in a condition suitable to picking up information, will be violated. This condition will be violated in a different manner if modifications or untypical conditions in the environment obtain, to the effect that the relations involved are not within the range of what is detectable even for a properly functioning perceptual system despite being nomological in kind.

Perceiving organisms, being finite creatures with finite resources acting within concrete ecological situations, are *in practice* not in a position to track all those specifying relations. Gibson himself admits that information may be inadequate, impoverished or masked in a given perceptual situation, so that conditions for perception fall out of the range of variance that an organism is accustomed to (Gibson 1966, Chapter XIV). Such will be the case when unusual lighting conditions obtain or when distorting mirrors or similar devices are placed in the environment. These qualifications, however, apply to what the perceiving organism is in a position to pick up from his environment rather than to the specificity of information as such.

In developing this account of "misrepresentation", Dretske (1986) seeks to accommodate for the possibility that an organism might get things wrong despite informational relations being in proper shape. Under normal conditions, he receives information on some world affair through several independent channels, and will be able to evaluate some signal or channel condition and remedy against what a faulty, information-less signal might seem to convey. Although he carefully avoids information-theoretical talk of "channels", Gibson apparently holds a similar view when he refers to information as being redundantly available in the environment, allowing for "multiple specification" (as paraphrased by Runeson 1988, 296f). Still, Gibson can be seen struggling with the concept of misperception (see Gibson 1979, 243f). He accepts that misinformation could still be information, while he insists that one will need separate theories of successful and unsuccessful perception respectively (Gibson 1966, 287f), so the parallel to Dretske's (1986) view that a theory of information cannot be symmetrically applied to cases of misinformation is only partial. Hence, the question arises of how to account for illusion and misperception if information is an objective commodity and perception is the activity of tracking the relations that make up that commodity.

Perceptual illusions vs. misperception: the empirical strategy

On some accounts, the possibility of misperception and illusion appears to be a hard, and possibly insurmountable, problem for an ecological theory of perception (see, first and foremost, Fodor and Pylyshyn 1981, 153–155). A short answer to this apparent problem is offered in a reply to Fodor and Pylyshyn from the camp of ecological psychology. It says that the Gibsonian conception of information "is roughly the claim that real possibilities are specified by current states of affairs" (Turvey et al. 1981, 293). On this view (which is largely shared here), the requisite information is present only if the affordance or "possibility" in question is real, and hence is warranted by current states of affairs. Sometimes, the perceiving organism will have to find out the hard way whether a possibility is real, and whether the information he seemed to pick up is there at all. Quite often, however, he has means of reliably making that distinction in the course of perceptual activity.

There is a variety of ways of being mistaken about some world affair, which are only partly acknowledged by Gibson. In his *Ecological Approach*, he only briefly and tangentially refers to misperception and perceptual illusions, and he does so primarily in the context of artfully created illusions. His prime examples are of two kinds:

(*pi* 1) Pictures that are purposefully made to create the appearance of objects that are not present in the environment.

(Gibson 1979, 281–283)

It is quite remarkable to hence find pictures subsumed under illusions, but this claim is consistent with Gibson's theory, and its normative connotations are not negative by default. A picture of some concrete object will contain *some* of the information that would be necessary for the object in question to provide the requisite affordance in its proper environment to its perceiver, yet without actually providing that affordance – for example sitting-on and sitting-at for the picture of a chair and a table. In such cases, we normally recognise both that would-be affordance and the pictorial nature of what is presented to us, and hence are not misguided. We are able to do so primarily because we actively explore what we perceive, by moving our eyes and heads in relation to an object. Only if and when we did not hence "test for the reality" of what we perceive, there could be such a thing as a perfect picture that tricked us into mistaking it for the scene that is being pictured (Gibson 1970, 426):[4]

(*pi* 2) Devices that are purposefully placed in the environment so as to create discontinuities in perception.

Drawing on the experimental work of his wife, Eleanor J. Gibson (Gibson and Walk 1960), Gibson cites as the paradigm of such artfully created illusions planes of solid glass that extend over visual cliffs and thus provide support while maintaining the visually based affordance of falling off the cliff (Gibson 1979, 142f). Here, conditions are modified in such a way as to create a mismatch between current states of affairs and an affordance or "possibility", which thus becomes "unreal". Misguiding the subject's perception is one central aim of the experiment, so the purpose of the artefact is clearly at variance with (*pi* 1).

These two cases, being quite distinct affairs themselves, are rather different, but not clearly distinguished by Gibson, from two other kinds of cases:

(*pi* 3) Naturally occurring perceptual illusions, such as illusions of length, colour or brightness.

(*pi* 4) Instances of misperceiving an object for another object that would have afforded different activities to the perceiver.

The primary issue here is not the naturalness vs. artificiality of the illusions and misperceptions in question – although, quite obviously, (*pi* 1 and *pi* 2) are based on artefacts, whereas illusions of kind (*pi* 3) occur naturally (while typically being investigated in laboratory settings), and misperceptions of kind (*pi* 4) might do so as well. More fundamental than this distinction, and not fully coextensive with it, is the one between perceptual illusions and misperception and their normative status.

In distinguishing illusion from misperception, I am following a fairly straightforward and presumably commonsensical distinction discussed by David Armstrong (1960, 4) in his interpretation of George Berkeley's theory of vision in Berkeley (1709): if I see something as red, round and having all the visual qualities of a tomato, and it turns out not to be a tomato but a plastic replica, I am likely to have misperceived it, for having mistaken it for another thing on the grounds of some similarity in perceivable surface properties. Still, I have not fallen for a visual illusion – that is unless I was mistaken about the replica's redness, roundness and so forth to begin with. If, in turn, conditions in the environment or my sense organs are such that I see something as square, purple and perhaps lacking other visual qualities of a tomato, and it turns out to be a very normally shaped and coloured fruit of that kind, I have been subject to visual illusion or hallucination respectively, in not getting the surface properties of the object right.

In either case, I might be at a disadvantage, and it seems natural to assume that this is the standard, the statistically normal result of instances of misperception, hallucination and illusion. If, on Gibson's theory, information for perception is supposed to specify an environment for a perceiving organism, it seems self-suggesting to characterise any divergence between what is perceived and how the environment stands as a failure, and to look either for deficiencies and dysfunctions in the organism's perceptual system or for inconducive or even treacherous conditions in the environment as the causes of that failure. And this is what Gibson does in his earlier *The Senses Considered as Perceptual Systems* (1966), and in his discussion of the difference between natural perception and hallucination (1970).

However, not every divergence between instances of perception and how the world stands needs to be a failure. The case is clear for hallucination but not for perceptual illusions. Perceptual illusions do not automatically result in misperception, and they may not be maladaptive per se – as Gibson's discussion of pictures as illusions that nonetheless preserve some of the information pertaining to their subject matter should have demonstrated. In naturally occurring perceptual illusions, the normative qualities of my perceptual relation are determined by the actual match or mismatch between the information that is present and the successful realisation of my abilities to act upon conditions in my environment, not the illusionary appearances. Getting all the physical properties of an object right might not be all that important in this respect (properties that Gibson did not care much about either). This is the basic idea brought forward by one relatively recent theory in cognitive psychology that has come to be known as the "Empirical Strategy" or, less memorably but more descriptively, the "wholly empirical approach to perception" (Purves et al. 2001; Purves et al. 2011).[5]

In combining an environment-directed outlook with a distinctive set of computational and statistical methods, the Empirical Strategy seeks to explain the peculiarities of perception by reference to a history of interactions between organisms and conditions in their environments. The Empirical Strategy is termed "empirical" precisely for rooting the character of perceptions in past experience of the individual or the species. How something *is perceived*, out of a spectrum of variant possibilities, is determined by how it *has been acted upon* – and not merely on how it has been perceived – in the past. Past success in doing something in response to a perception will act as a necessary condition in determining how the object or scene in question is being perceived at present.

The authors commence from the "inverse optics problem", also known as the problem of the ambiguity of the retinal image, as it was paradigmatically formulated in Berkeley's *Essay Towards a New Theory of Vision* (1709) and briefly discussed earlier in this chapter. In view of this problem of perceptual ambiguity, Berkeley's proposal was to ground the ability to perceive distances and magnitudes of objects and, in fact, any spatial arrangement, in the perceiving subject's experience. Only from experiencing certain objects in certain constellations, one learns to associate prima facie identical visual cues with different perceptual situations, and hence to reliably identify the correct distance, magnitude, and so forth of objects.

Purves and his colleagues follow Berkeley's lead when identifying the inverse optics problem as the issue "that light stimuli cannot specify the objects and conditions in the world that caused them" (Purves et al. 2011, 15588), and, like Berkeley, they delegate the task of specification to the perceiving subject's experience. Their solution to this problem, however, is not as exclusively based on subjective experience as is Berkeley's, and it certainly does not follow Berkeley's idealism. Instead, they adopt a more externalist and supra-individual stance. Conversely, with Gibsonian ecological psychology, the Empirical Strategists share an emphasis on the relation between perception and environmentally embedded behaviour and a markedly realist outlook.

The core of the Empirical Strategists' argument can be parsed into three steps, and it is this: "Proximal stimuli trigger patterns of neuronal activity that have been shaped solely by the past consequences of visually guided behaviour" (Purves et al. 2001, 285). These consequences, in turn, are evaluated purely in terms of adaptive success rather than any correspondence with the measurable physical properties of the perceived objects. Hence, the measurable physical properties of some object provide only a set of boundary conditions that underdetermine the possible ways of perception. At any rate, they are not represented in vision (Purves et al. 2011, 15592).

In this circumscribed sense, the Empirical Strategists share Gibson's view that perception is not a process of representing the physical properties of some object or scene but an interaction with some concrete object in its concrete context under concrete conditions of perception. What one can do with or about that object in a given situation or, in Gibsonian terms, what that object affords to the organism is what fixes the way in which it is perceived.

As conceived under the Empirical Strategy, the relation between perceptual qualities and their target appears at once rather simple and fairly complex and abstract: some perceptual token will be reproduced if it is closely associated with successful behaviours towards its source or towards some correlate of that source. This condition is sufficient for a very specific sort of empirical adequacy. This empirical adequacy is very specific because it does not build upon any direct relation – and perhaps not even a proper covariance – between a perceptual quality and the physical properties of the perceived object. Divergence between perceptual qualities and the physical conditions at the source does not amount to misperception: "Since the measured properties of objects are not perceived, they cannot be misperceived" (Purves et al. 2001, 296).

Instead, what is decisive for getting things right in perception are the frequency distributions of different retinal patterns, which are mapped *not* onto variance in physical variables *but* onto variance in behaviours that differentially respond to certain world affairs, in accordance with the probability distributions of the occurrence of these affairs (Purves et al. 2011, 15594). For example the observable mismatch between differences in lightness or brightness of an object as perceptual qualities on the one hand and measured illumination and luminance of the physical objects on the other is attributed to the relation between two factors:

(*f*1) Frequency distributions can be determined for variant luminance values of some object as they obtain for the contexts of the various natural scenes in which it, individually or as a member of a type, appears. Objects of some kind will be more often encountered under certain lighting conditions than under others.

(*f*2) Frequency distributions are assumed for the rates of success of behaviours of the perceiving organism or his ancestors towards that kind of object under variant conditions. These frequency distributions will be affected by processes of selection, on phylogenetic or ontogenetic levels, of variant behaviours.

These two types of frequency distributions can be mapped onto each other, so as to see how reliable behavioural success with respect to the object in question will be under the predominant conditions of appearance in the perceiving organism's environment, and what the cost of failure under less frequent conditions will be. The general strategy is to match perceived qualities of some object against databases of frequencies of occurrence of retinal images corresponding to commonly occurring natural scenes containing that kind of object.

In the present example, equally luminous objects or patterns are perceived as darker when placed in a brightly illuminated context and lighter when placed in a darker context because the differential rates of occurrence of the retinal projections caused by the same objects under lighter vs. darker conditions are matched by adaptive behaviours towards those same objects under the respective conditions (Purves et al. 2011, 15589). At first, the probability distributions of luminance values of objects under lighter vs. darker conditions and the perceived brightness will seem to be skewed towards greater perceived brightness than measurement of luminance would suggest for those retinal patterns which occur most often, namely under poor lighting conditions. However, perception is *not* skewed in terms of the behaviours that respond to the world affairs as they are encountered under these conditions. There will be some use to perceiving objects as exceedingly bright under poor lighting conditions. The use of this seeming illusion should be expected to lie in more reliable or efficient recognition of the kind of object in question when lighting conditions are poor, and hence in more reliable responses to the presence or behaviour of that object. Similar conditions apply to other seeming illusions, such as illusions of length or colour.

Alternatively (a possibility not discussed by the Empirical Strategists), there might be cases where a seeming illusion is a side-effect without adaptive functions of its own (or, in the terms of Gould and Lewontin 1979, a "spandrel") of perceptual processes whose mechanisms of production serve some adaptive purpose. Such side effects would then either be adaptively neutral or only mildly detrimental – so mildly in fact that they are more than outbalanced by the benefits gained from the successful performance of the adaptive functions of the mechanisms that happen to generate them. Under these latter conditions, getting the luminance values wrong would be the artefact of the operations of mechanisms of perception whose success is more relevant to the organism in positive terms than the bias in perception of luminance is in negative terms. But even then, there would have to be some kind of correlation between the frequency distributions, across the perceptual situations the organism encounters, of the physical variable under observation ($f1$) and the presence of some other factor that is relevant to the rates of success ($f2$) of those behaviours guided by that other mechanism. Otherwise, no statistical relation between $f1$ and $f2$ could be identified to begin with.

Either way, the normative implications of getting the luminance values or other physical variables wrong are either inverted or at least much relativised. The length, shape, colour and so forth of some object may be invariably misrendered in perception, hence misperceived, while being appropriately treated on the level of behaviours and their adaptive functions. If the effects of the behaviours towards the object in question are conducive to the perceiving organism's welfare, in the light of his overall constitution and abilities, and if this conduciveness is rooted in a history of phylogenetic or ontogenetic selection of whatever ways of perceptual rendering of the respective properties may come to pass, that perceptual rendering will be vindicated. Systematically misperceiving the physical properties of some object, in terms of measurable variables, might be essential to correctly perceiving the object in terms of what can and should be done with it. Hence, the apparent mismatch between experimentally determined perceptual qualities and measured physical variables is an epiphenomenon of the operations of properly functioning mechanisms and may be discounted as such.

Ecological psychologists after Gibson have offered partly similar accounts of the relation between illusion and misperception. With reference to the Ames's distorted room illusion, Sverker Runeson (1988) argues that prima facie illusions are easily corrected for in the course of perceptual activity, and that, in order to persist, illusions typically have to be meticulously created and maintained – which usually does not occur under natural circumstances. Perceptual ambiguity in terms of equivalent configurations in the optic array may be geometrically possible, but these configurations are either physically impossible or encountered only in manipulated environments. Either way, Runeson argues, these equivalent configurations would be informationally irrelevant. Moreover, any residual ambiguity that might get in the way of correctly perceiving affordances is likely to be mended in the further course of perceptual investigation, and in developing one's perceptual skills.

Perceptual learning might allow for correct judgement about some state of affairs even when an illusion persists, argue John Kennedy and colleagues with respect to geometric illusions (Kennedy et al. 1992): When presented with two well-known size illusions (the Jastrow curves and the Sander parallelogram), transformations of the respective shapes that correspond to perceptual investigation of an object from different angles allowed the majority of the subjects in the experiments to make correct judgements about the true size ratios between the shapes – without the size illusion actually being dispelled. One shape still looked larger than the other but now was known to be of the same size. Although Kennedy et al. (1992) do not make that inference, one should expect that an affordance related to the objects in question would have come to be correctly perceived despite the persistence of the visual illusion.

An even more direct match between an ecological view and the argument of the Empirical Strategists is provided by Qin Zhu and Geoffrey Bingham (2011): one of the most robust natural perceptual illusions is the size-weight illusion, in which an object will be perceived as being heavier if and when it is smaller than another object of equal mass. That illusion, the authors seek to demonstrate in an experiment, has a correlate in human subjects' learning of perceiving throwing affordances in terms of selecting objects for optimal size-weight ratios. These ratios are correctly chosen by proxy of a biased perceived quality. On the background of an evolutionary argument, it is concluded that the illusion has a function in terms of guiding human beings to pick objects that are optimally throw-able over long distances, which was a highly relevant skill in Palaeolithic hunter-gatherer societies, and which has been identified as an ability as uniquely human as language.

In light of these observations on the possible adaptive functions of mismatches between measured physical variables and perceptual qualities, should these mismatches still count as instances of misinformation? Given the relation between the perception of affordances and the notion of natural information that I suggested was held by Gibson, there might be a good reason for believing otherwise: to the extent that an external observer could apply point-to-point mappings, in terms of a multidimensional mathematical function between, first, the variable as it is being registered in any perceptual situation that falls within the range of what the perceiving organism is adapted to, second, the variable as it can be measured by that external observer, third, the concrete conditions under which the ability to perceive that variable in a certain way is realised and, fourth, the adaptive function of that ability under the given type of conditions, the information that is seemingly misrendered is actually preserved.

Any apparent mismatch between perceived qualities and measurable variables detectable in perception can be resolved by accounting for the very situation in which perception occurs and the adaptive function of the mapping that applies to this type of perceptual situation. If, under certain conditions, something looks longer, brighter and so forth to the organism than measurement would suggest, these conditions and the functions which that seeming departure serves override correspondence with physical variables. More precisely, adding a "condition" and a "function" variable to the equation will keep the informational mappings intact and allow for an, albeit complex and indirect, correspondence to the values of the physical variables – if and when perception works normally.

Under a genuinely ecological perspective, and in keeping with the strategy of the Empirical Strategy, perceived qualities will be overridden by the informational relations on which successful activities are grounded. Genuine misperception will occur only if, returning to Warren's aforementioned case, the subject perceives a stair to be climbable when it is, in fact, not climbable, or vice versa, or if he mistakes an exhaustingly steep stairway for a conveniently climbable one.[6] The subject will then misperceive the respective affordance – which does not even need to imply that he or she is mistaken about the absolute physical measures of the object. Conversely, perceptual illusion will occur if the subject turns out to be mistaken about absolute, extrinsically measured riser heights or transformations of heights in the course of the experiment – which, if the experimental set-up matches normal conditions for perception, is unlikely to negatively affect perception of an affordance. Hence, perceptual illusion and misperception of affordances may coincide but are independent issues. In some cases, perceptual illusions may even contribute to the correct perception of an affordance.

Although the preceding discussion suggests that the Empirical Strategy in some respects elegantly complements the ecological view, two notable limitations to this approach should be mentioned here: first, almost all examples discussed by the authors are concerned with cases in which identical targets look different to the perceiver under different conditions, whereas their point of departure, the inverse projection problem, was identified by Berkeley as the problem that different targets, when placed in a certain relation to the perceiver, will look identical. Berkeleyan ambiguity does not imply that perception is *at variance* with the measurable properties but that it becomes ambiguous by *one* retinal image being in accordance with *different* instances of measurable properties. The Empirical

Strategy thus accounts for only one of two types of cases of illusion (see Purves et al. 2011, 15590). Second, the Empirical Strategy remains exclusively concerned with isolated perceptual qualities, such as brightness or perceived angles, length and motion rather than more complex perceptual affairs in which seeming illusions are mended in the course of perceptual investigation, and in which an object with complex and perhaps multi-modal perceptual qualities is relevant to the organism in certain ways. In contrast, Gibsonian ecological psychology refers to the environmental context and the additional cues it provides to the organism, so as to correct for perceptual ambiguities of the second, the Berkeleyan kind.

Despite these limitations, the Empirical Strategy is correct in suggesting that a perceptual illusion, in terms of a prima facie mismatch between perceptual quality and measurable values for some physical variable, may serve adaptive functions. A certain class of perceptual illusions, one that is not clearly distinguished from instances of misperception by Gibson, can thus be accounted for. A necessary precondition for the accomplishment of the adaptive function of a perceptual illusion will be that the illusion is embedded in the context of an environment that is stable enough to allow the organism to handle world affairs with some reliability and within the bounds of his constitution and abilities. Genuine misperception occurs only if and when either the conditions within organism or environment depart from what he is adapted or accustomed to or if and when the perceiving organism fails to grasp what can or should be done with the object in question. Perceptual illusions, if and when they have acquired an adaptive function, may actually both be a constituent of the perception of affordances and provide accurate information about some world affair to the perceiving organism. Under normal circumstances, perceptual illusions of the kinds discussed previously, for their very regularity and their possible adaptive function, have precious little to do with the misperception of affordances but, very much to the contrary, with helping the organism to get right about some object what he *needs* to get right.

Any insect replica in an experiment in frog vision that shares most or all perceivable (i.e. 'frog-perceivable') surface properties with edible insects will be perceived and treated identically, however, without being nourishing to the frog. Such ambiguity can be resolved only if other cues as to the insect replica's identity as a replica are or become available to the frog. These are situations of genuinely ambiguous information for perception that is prone to end in misperception – but misperceptions, so understood, are certainly not cases of perceptual illusion of any kind. Misperceptions of this kind will be difficult to assess for any theory of perception that does not spell out the specific conditions for perception with respect both to its historical *and* to its ecological context.

If, in turn, studies in the measurement of affordances are right in assuming that the information relevant to perception normally is intrinsic, the perceiving organism's correct grasp of his relation to the environment will override any apparent mismatch between perceptual qualities and the absolute values for some physical variable that could be measured by an external observer. Misperception, then, can only be misperception of objects in their entirety, not of their measured properties. In Gibson's terms, such misperceptions would be misperceptions of affordances – not, for example, misperceptions of spatial relations or colours. After all, perception, even if it can be assessed in terms of measurement of physical variables by an external observer, does not amount to such measurement *by* and *for* the perceiving organism.

This observation will fit rather well with Gibson's contention that we do not perceive such abstract relations as space – a contention that grounds his ecological approach (Gibson 1979, 32). It is the *ecological* environment rather than an abstractly and formally described physical space that is directly and objectively given to the organism, and that he acts upon. He is perfectly entitled to get things wrong in terms of conditions in physical space as long as he gets them right in terms of the ecology of perception. Still, the ecological environment, despite being objective in nature, cannot be defined in abstraction from the organism inhabiting it. He contributes to the informational relations that he uses. His contributions, if regular in form and adaptive in function, will neither make that information a subjective affair, nor give rise to the misperception of affordances.

Notes

1 In a revealing passage, Marr (2010, 29) criticises his predecessor and opponent for that "he did not understand properly what information processing was, which led him to seriously underestimate the complexity of information-processing problems involved in vision". If one is convinced that information is not processed in perception in the first place, one will hardly feel exposed to the charge of underestimating the complexity of that processing.

2 The mediation I am alluding to was through the work of Donald Norman (2002), who applied a rather liberal interpretation to some of Gibson's core tenets.

3 There is no unequivocal definition of information in Gibson's 1966 and 1979 books, nor is there one in his smaller works on information (1960; 1971).

4 Several years prior to the *Ecological Approach*, Gibson was involved in a debate with Ernst Gombrich and Rudolf Arnheim on the nature of pictorial representation, in which the perceptual status of what is pictured in a picture and its recognition was the topic; see Gibson (1971) and Gombrich et al. (1971). My acknowledgements go out to Alfred Nordmann for highlighting the subtleties of this debate.

5 I have to thank Brian McLaughlin for introducing me and others to the Empirical Strategy during the workshop "Perception and Knowledge" at the University of Graz, Austria, in October 2012, and I have to thank Martina Fürst and Guido Melchior, the organisers, for placing that workshop and McLaughlin's paper right when and where I needed it.

6 In fact, as Warren (1984, 695) reports, a stairway can be just as exhaustingly gentle, with low raisers and deep treads, as it can be exhaustingly steep. The perceived gentleness of ascent may indeed be a common misperception of an affordance.

4 The domains of natural information

If natural information is to be properly understood as something provided by the environment to its consumers, and hence as something that grounds their perception and behaviour in the first place, the discussion in the preceding chapters will have helped to identify some of the properties to be expected from this kind of information: Natural information shall be an objective commodity but also allow for the possibility of being relevant in different ways to different consumers. It specifies the environment for the perceiving organism but often leaves the organism with situations of underspecification. It may convey knowledge but most of the time serves more immediate uses, as in Gibsonian affordances. It shall provide for the possibility of getting things wrong but also help to account for how we can get them right. It shall not presuppose the presence of beliefs, desires and knowledge but be part of their causal-historical preconditions. It shall be well-grounded but need not be symbolic. And it shall allow for accommodating changing conditions, over place and time, in the environment. An understanding of these properties of natural information will help us to understand just how an organism's environment contributes to his cognitive accomplishments (further discussion in Chapter 6), and what to look for when considering the role of information-providing artefacts in this context (further discussion in Part II).

What I have not provided so far is my own account of natural information, its rooting in the environment and its role to its consumers in precise enough terms to add something worthwhile to the earlier attempts discussed in Chapter 2: Fred Dretske presupposes that information should be natural information in the first place, and he provides a definition of informational content that is firmly rooted in the laws of nature but finds it difficult to accommodate for the conditions of information uptake in the real world. Ruth Millikan criticises Dretske for restricting the presence information to a small subset of real-world conditions, maintaining that organisms typically rely on more local and probabilistic conditions as information. Brian Skyrms's and Peter Godfrey-Smith's Lewisian approach, in turn, while being ecumenical about the criterion of lawfulness, are at risk of remaining in the grip of game-theoretic metaphors that presuppose active senders and receivers of information. Among the theories of perception discussed in the preceding chapter, James Jerome Gibson's approach comprises a notion of natural information that, once made more explicit and precise, might help to carve out a synthesis of those views, which to undertake is the purpose of this chapter. I will outline the concepts of natural information and of informational domains that, I suggest, are the best way to go, and then very briefly recapitulate the lessons to be learned from Dretske's and Gibson's notions of information.

Natural information and reference classes

Let me try to provide some theoretical definitions that, although certainly not aspiring to constitute a fully fledged theory of natural information in its own right, or to provide a genuine conceptual analysis of "natural information", still integrate and synthesise the insights from the preceding discussion in a way that describes natural information as both objective and context-sensitive enough to be useful to its consumers under the manifold conditions under which it is encountered. Above all, it shall help telling apart the variant from the invariant aspects of what a perceiving organism encounters in his environment. The basic idea to be promoted here is that natural information is both a nomologically governed relation *and* domain specific in relevant respects.

(NI-1) The *invariance* condition: informational relations obtain if and only if distal conditions *F* and their transformations at *s* are matched by detectable proximal conditions and transformations at *r* in accordance with a strict natural regularity *I* that requires a uniform explanation.

(NI-2) The *reference class* condition: the probability of a certain informational relation between *r* and *s* being *F* to obtain in accordance with regularity *I* is determined with respect to a certain reference class *C*, where

- (a) *C* may universally comprise all real and possible situations of *I*-regular covariance between type-**r** signals and *s*-conditions – the upper boundary case;
- (b) *C* may comprise a spatio-temporally bounded region in which type-**r** signals covary with *F*-conditions at *s* under *I*, with other conditions concerning *r*, *s* or *F* obtaining outside that region – the case of *locally* bounded information;
- (c) *C* may comprise a regionally bounded or global set of type-**r** signals of which only a *subset* $\{r_1^i, \ldots, r_n^i\}$ covaries with *s*-conditions under *I*, with no means of further discrimination available – the case of *probabilistically* bounded information.

(NI-3) The *receiver* condition: the definition of the reference class *C* is dependent on the concrete situation, purposes, constitution and abilities of a receiver *R*.

Part NI-1 should be general and basic enough to capture Dretske's central claims on the nature of information discussed in Chapter 2 while omitting his reference to knowledge and conditional probabilities. Only in NI-2, do receivers and probabilities begin to make an explicit appearance. In the case of NI-2a, the receiver will be an external observer by default, whereas in cases NI-2b and NI-2c, the receiver will typically be involved in a concrete situation of tracking information that is relevant to him. Conversely, what is relevant to him and what he is capable of tracking will make a difference to the probabilities of the signals he uses to actually match distal conditions. The issue of the level on which and the degree to which the receivers or "consumers" of natural information become important is hence related to the issue of the degree to which natural information may be an openly probabilistic affair, that is an affair allowing for probabilities $p < 1$. These two points of contention concern the proper choice of a framework of reference for determining the probabilities of informational relations (see p. 23 in Chapter 2).

Already quite early in the discussion of Dretske's (1981) information-theoretical account of knowledge, Gilbert Harman (1983) pointed out that one could determine the probability of a certain outcome in any information-generating situation only in relation to a given "framework", so as to measure the reduction of possibilities that takes place – a point not contested by Dretske (1983, in his "Author's Response" to Harman's peer commentary). The observation of a coin-tossing game with a set of unequally fair coins, for the mixed probabilities involved, will deliver different values than the observation of frequencies of results for each coin individually. Similarly, Dretske's requirement of $p = 1$ for informational content can be met only if the framework or "reference class", as Dretske himself calls it, is tailored tight enough to allow only this outcome, so that, whatever the initial distribution of possible states has been, the result is unity. The observation of a coin-tossing game, if comprising a reasonably large sample and if confined to one or the set of all fair coins, will never produce a conditional probability $p > .5$ over the course of that observation. Here, the reference class is the entire game or set of games, as observed over a number of rounds between t_0 and t_n. However, quite obviously, we will know with certainty what result only *one* toss in the series produces after we have observed that one toss of the coin. Here, the reference class will be restricted to one single case, where this case, in turn, is supposedly governed by laws of nature – that set of laws which is sufficient to explain why the coin, under the given set of antecedent conditions, landed this way.

Cases of this latter sort are what Dretske envisions. Only under these conditions, Dretske's underlying argument goes, information may convey knowledge. There would be no knowledge if conditions at the source were allowed to be at variance with what is being signalled with whatever small degree of likeliness. On these grounds, he restricts the assignment of the status of information to those cases where the condition of unity is met. Probabilities concerning the entirety of conditions within sets of events may be shaped by a variety of factors, and are ultimately not his concern.

As authors such as Millikan and Skyrms argue with some justification, setting admission criteria to the realm of information as strict as Dretske might help us to an account of information that satisfies a (foundationalist) criterion of knowledge but bears little information on how organisms successfully perceive and act under less tightly bounded but more natural conditions. Knowledge may not always and under all circumstances belong to the conditions of success in perceiving and acting within an environment. Conversely, setting the admission criteria too loose might be similarly dissatisfying, because there are innumerable ways in which all sorts of world affairs are related, with whatever degree of probability. If such relations could pass as informational without being reasonably unambiguous and reliable, and perhaps even without being detectable to any organism, the cognitive value of information would be inflated away.

The previous definitions, in their reference to strict natural regularities and reference classes, aim at allocating a proper place both to the "nomological" or "specification" views defended by Dretske and Gibson and the "probabilistic" or "local" views advocated by Millikan and Skyrms in particular. The mark of distinction between the respective domains of application of those accounts is that between the description of information as being *present* in the environment (NI-1) and as being factually *available* to a receiver (NI-2 and NI-3). The claim that the limited and imperfect availability of natural information to an organism leads to underspecification, as indicated in NI-2b and NI-2c, does not militate against the specificity criterion in NI-1.

According to NI-1, information basically *is* specification of the environment. It is so to the extent that there are invariant relations between world affairs that could be tracked by some receiver, if only hypothetically. If there were no such relations, a receiver would have no possibility of keeping invariant aspects of the environment apart from the variant ones. Nor could he keep rule-governed, invariant-preserving transformations apart from random ones. On this level of analysis, the reference class for the informational relations will comprise of the set of all uniformly regular invariant relations in the world that could be possibly tracked by an – unspecified – observer. Hence, this part of the definition covers both Dretske's requirement of a nomologically governed probability of $p = 1$ for information to obtain and the independence of that relation from any particular observer. By implication, NI-1 excludes from the class of informational relations such relations between world affairs that are probabilistic, for being governed by a disjunct set of regularities $\{i_1, \ldots, i_n\}$ or for including an element of randomness and thus altogether failing to meet condition *I*.

Conversely, NI-2 excludes the hypothetical case of relations that, even if they were properly regular, would be unobservable in principle (although perhaps manifesting themselves in other, intractable ways). Informationally inscrutable situations in this sense are part of certain metaphysical or religious doctrines, such as the Hegelean *Weltgeist*, cosmic conspiracies, the Judeo-Christian inscrutable God or the Occultation of the Mahdi in Shia Islam. For these sorts of boundary case, there would be no reference class identifiable in accordance with NI-2 or any of its sub-clauses.

Epistemically more mundane and relevant are situations of epistemic uncertainty as they are characteristic of many areas of probabilistic, statistics-reliant science. In this kind of case, one typically expects some natural regularity or law *I* to underlie a set of variant phenomena, while having to rely on signals that are not unequivocal enough to determine that regularity, or *r* as an expression of that regularity with certainty. Intervening factors will be held accountable for blurring the signal but are either of random nature or too many and too complex to actually be accounted for. What can be achieved instead is partial, approximative confirmation or refutation of a hypothesis within the bounds of available data and applicable methods. NI-2c covers this sort of case in particular, with the possibility that NI-2b may apply, too.

For example there is the possibility that someone contracts an infectious disease, let us call it *u*-fever, while lacking the key symptoms of the illness, which might be called *u*-rashes, or even remaining asymptomatic. The condition of *s* being *F* will then be fulfilled, and *I*-regularly so, but *r* does not always obtain, so that false negatives obtain. This is a frequent case in medical practice, and it belongs to the realm of possibilities that, like the possibility of indiscernible symptoms with variant aetiologies, educate any responsible physician to being a careful probabilistic reasoner.[1] If *u*-rashes could be caused either by *u*-fever or, say, morbus *y*, that symptom by itself would not provide information on the underlying condition within the context of the reference class relevant to the physician. If, however, the *u*-rashes symptom occurs only intermittently but in rather exclusive connection to *u*-fever, and if the proper aetiology of that symptom is in place, and if the intermittency of the symptom can be ascribed in contextual conditions, *u*-rashes will count as a signal of *u*-fever, and hence provide natural information about the presence of *u*-fever in the patient. With respect to intermittency, the uptake of additional information about *s* by *R*, taken up by different means, will help to reduce uncertainty on the receiver's side as to whether the illness is present in the absence of that typical signal.

However, most relevant to the present context are situations of underspecification that occur in perception in a natural environment. On the one hand, conditions in the ambient optic array, to recur to Gibson's terminology, are specific in terms of the laws of reflection, refraction and spatial geometry being in place. On the other hand, visual information may become masked, blurred or otherwise distorted, or conditions are such that some of the information present will not be taken up at all. This is the case for NI-2c in perception in natural environments. Alternatively, some of the information may be reliable only in certain places, at certain times. This is the case for NI-2b in perception in natural environments. In reversed order of argument to the previous set of examples, NI-2b is the paradigm case of underspecification in natural perception, where NI-2c will often also apply. The reference class for the signals involved will not comprise probabilities near the Dretskean condition of unity in NI-2c cases. Under NI-2b, they might but need not locally approach it.

In view of this issue, and along the lines of Millikan's critique of Dretske, Matthieu de Wit et al. (2015), together with Rob Withagen and Anthony Chemero (2009) suggest a 'softer' notion of natural information as the biologically more realistic one: "epistemic contact" with the environment, in Withagen and Chemero's wording, typically comes in degrees. Perfect contact, that is full specification, is practically unattainable for organisms under natural conditions of perception, and hence will pose too high a requirement. Given that organisms are always bound to locally and temporally variable environmental conditions, this argument has some prima facie plausibility. As de Wit et al. (2015) admit, locating ambiguities in perception in the informational relations themselves will be a concession to "inferential" approaches to perception, for the sake of ecological credibility. However, achieving such credibility does not require one to accept the inferentialist "doctrine of intractable nonspecificity" of informational relations (as Turvey and Shaw 1979 call it). If informational relations may hold between all sorts of world affairs, and if they are regular in the way outlined by Dretske, there will be relations in the environment that remain fully specific even under changeable ecological conditions.

Although highlighting the importance of the degrees of a perceiving organism's epistemic contact with his environment is well-taken, acknowledging situations of underspecification does not require one to admit for a nonspecificity of information. Just as the presence of invariants in the environment does not imply invariance in ecological conditions for perception, variance in ecological conditions for perception does not amount to a weakening of invariant relations in the environment. Indeed, camouflaged predators indiscernible from harmless twigs may evolve, or humans may invent windows that happen to block a bird's flight path. In such cases, the optical invariants that are amenable to perception for the animal and that had hitherto been sufficient for specifying to him an ecologically relevant condition in general or an affordance in particular remain in place, whereas some of the distal conditions typically related to these optical invariants have changed for a relevant subset of cases. However, if an optical invariant had been sufficient to specify an open flight path to the bird before the invention of glass windows but now is insufficient to do so, that invariant itself will not have become compromised after that invention. What has happened is that this optical invariant alone cannot be used as a reliable proximal signal of the distal ecological condition anymore, and hence is now associated with a probability $p < 1$ within the reference class of the type of proximal signals available to the bird.

Ecological changes affecting epistemic contact do not *per se* constitute a problem, yet there will be a penalty for being mistaken too often if and when the receiver depends on discriminating between different conditions with a certain degree of reliability. It is the requisite degree of reliability that determines the reference class for the signals and their probabilities. If the first and slightest mistake is sufficient to get the organism into trouble, the reference class will be restricted in such a way as to approximate Dretskean rigidity. If an individual or population can live with 95% false alarms, conditions will be much more relaxed. Still, all rounds of picking up the putative signal will be counted into the respective reference class. An organism is not typically in a position to gather information about changes in ecological conditions or the degree of reliability of a signal. Depending on where the acceptable margin of error lies, he will either have to live with increased uncertainty or undergo some adaptive reorganisation. He must then either be or become able to detect where and when the reliability of the signals attains specificity, which, under natural conditions for perception, seldom occurs, or he must detect where it remains high enough to meet his or her requirements.

If he has no such means of detection, he must remain confined to living within the boundaries of the region in which these conditions hold. I will discuss this point, introduced by Millikan as the "domains" of natural information, in more detail in the next section.

While NI-2 considers the question of specificity vs. probability of relations within a reference class as being dependent on the constitution and abilities of a receiver, and hence refers to NI-3, the receiver condition stated in this clause also caters for the intrinsic nature of natural information to a perceiving organism: informational relations may not always and under all circumstances involve a straightforward, organism-independent matching between a received signal and the values of the physical variables at the source. As the discussion of perceptual illusions in the previous chapter should have shown, a signal may be found to *never* match the value of a variable at the source and still be useful to a perceiving organism. Getting the values of some physical variables at the source wrong may, under certain circumstances, actually be in the service of getting something more important about that source right, namely what can or should be done with respect to conditions at that source. If an – external – observer of the informational relation limits his perspective to conditions at the source and types of environment-bound signals related to that source, the relevant pattern will not be properly detectable for him to begin with. Only to an external observer who does thus not account for the intrinsic, organism-involving mappings that are part of the informational relations, will there be an informational bias involved in perception. To the perceiving organism, the reference class of the signals may well comprise of tokens mapping distal conditions in indirect but highly reliable ways.

The correspondence to the physical variables does, however, not typically become explicit to the perceiving organism in the perceptual process. Otherwise, the perceptual tokens would have to carry a mark of how they map onto the physical variable under a defined range of conditions – something like an indicator of the complex affair "the object looks *n* degrees brighter now that the total amount of visible light decreased by *m* lm". Although there is no such indicator, the informational mappings will remain intact to the organism. If, in this kind of case, a bias in perceived qualities were to result in two different values for some physical variable being perceived equally, such as equal perceived brightness for variantly luminous objects, an ecological argument might be applicable (or even necessary) to account for such situations: the equivocation either does not matter, in that it does not, or not significantly, interfere with the attainment of the organism's goals, or the other type of object or lighting condition does, as a matter of empirical fact, rarely or never occur in *R*'s environment, where that matter of fact should be due to one or a number of natural regularities in their own right.

Hence, there is no "too bright" in the perception of some object if the appropriate perception of the conditions at the source – the affair to which the perceiving organism is supposed to respond and hence the affordance that it presents to him – depends on perceiving the object in question as being brighter than the measured luminance values would suggest. Although the organism might thus appear to be misinformed about the luminance values at the source, he will be correctly informed about the presence of the object that he will need to recognise in order to persist. By virtue of its regularity, the seeming mismatch will constitute an informational relation in its own right. Again by virtue of the regularity of these relations, the inclusion of organism-involving relations is perfectly compatible with the notion of information being an objective affair. It may be true that the informational relation in question would not exist in absence of the receiver, but only in the same trivial sense in which any informational relation depends on the existence of certain world affairs.

The regularities involved in this complex mapping relation will have been established in a history of interaction, on individual or population levels, with those states of affairs. The seeming departure from the values of the physical variables involved in such situations is not at all capricious but depends on processes in which that deviation acquired an adaptive function that consists precisely in enabling the organism to get a grasp of the conditions at the source that are relevant to him and his kind. The probabilistic element in natural information, on this level, is not one of random distributions in which, for example, an object appears n degrees brighter than measured at t_0, exactly as bright as measured at t_1, n degrees darker at t_2 and so forth, but in the frequency distribution of perceptual situations under which the function of perceiving objects of that kind in a certain way evolved – notwithstanding the possibility of false positives or false negatives in perception, and allowing for some degree of variability in perceived qualities over a series of perceptual events.

The preceding discussion should have served to show that it is the situation, the constitution, the abilities and the purposes of the receiving organism that are indispensable to the definition of the reference class for the information involved. I will try to articulate the upshot of that discussion, and add more detail to clause NI-3 in particular, by stating a set of conditions for the successful use of natural information by its receivers:

(NI-4) The *detectability* condition: different signals $\{r_1, \ldots, r_n\}$ must be discernible to some determinate degree for some receiver R, where

- (a) the conditions at s to be detected by R are determined with respect to the constitution and abilities of R – the condition of the *intrinsic definition of the reference class* for the mapping relations of $\{r_1, \ldots, r_n\}$;
- (b) the mapping relations between $\{r_1, \ldots, r_n\}$ and the conditions at the source make a difference to R in terms of the fulfilment of R's specific purposes – the condition of *relevance*;
- (c) the conditions for information uptake are such that signal tokens $\{r_1, \ldots, r_n\}$ occurring in accordance with the respective conditions at s remain detectable for R over time – the condition of *stability*;
- (d) the frequency for R of recording r_1 where s is not F, or where some other r_n occurs instead of r_1 or where no r at all occurs remains below a threshold that is critical to the fulfilment of R's purposes – the condition of *unequivocality*.

The conditions in NI-4c, as the "channel conditions", are based partly in the organism, partly in the environment. They include the organism's perceptual apparatus as well as enabling and limiting conditions in his surroundings for the detection of informational relations, and these, in turn, may include general physical conditions, such as lighting, as well as, more formally and generally, the level of noise and error that are present to interfere with information uptake. Nonetheless, neither the perceiver's condition nor the environment are supposed not to be capricious, that is the channel conditions are not supposed to change randomly.

At what level an ability to discriminate between different conditions at *s* is located and how many information channels are available in what quality will depend on what may be called the adaptive economics to which an organism is exposed. An organism who would have to spend a significant effort or a species that would have to invest a lot, in evolutionary terms, into an elaborate perceptual apparatus that could discern between, for example specimen of an edible and nourishing but relatively rare prey species and a more populous sort of nutritionally unrewarding but not poisonous lookalikes, where specimen of both kinds are equally easy to catch, will get along well with his inability to discern between *r*-signals correlated with the less numerous but nutritionally rich s_1-specimen and the more numerous but nutritionally poor s_2-specimen. The condition that determines the value of such trade-offs between the cost of an investment into the possibility of finer discrimination and the effort spent on a large number of useless catches will be the ability of the organism or species to maintain and reproduce themselves. From the perspective of the prey, a complementary situation arises: what is the trade-off between the cost of investment into an ability of reliably distinguishing between a rare but efficient predator and his numerous but rather harmless lookalikes or investing into an elaborate camouflage or mimicry that interferes with detectability for the predator on the one hand and the cost of reproduction of another specimen of one's own kind on the other?[2]

This sort of adaptive economics is supposed to apply not only to natural organisms but also to technological systems, minus the supposedly red-in-tooth-and-claw mechanisms of selection at work in the biological realm. In either type of case, there are different strategies for safeguarding against disruption: a refined apparatus designed to always record information as accurately as possible through as many channels as can be made available, or the design of choice may be one for resilience, by means of robustness or redundancy, against the effects of numerous misreadings. The seeming simplicity of the latter strategy is not to be confused with any sort of adaptive disadvantage. However, not all strategies will work equally well under all conditions. A complex but informationally relatively tractable environment is likely to suggest the former strategy.

If not earlier, as Skyrms and Godfrey-Smith suggest, it is at this point where game- theoretical considerations, if carefully hedged, become helpful to an inquiry into natural information and its use. The receiver of information need not be consciously engaged in applying an economic calculus to guide his behaviour. The relation among the information present, the number and quality of the information channels available to him, and the expected results of his behaviours will be sufficient. It is in this light that NI-4d and the condition of the "determinate degree" of discernibility in NI-4 should be understood. There is no awareness required from the receiver of the presence and kind of the informational relations that he relies on. It will be the privilege of the followers of the complexity-oriented strategies to develop some degree of awareness, perhaps even a conscious one, of the informational relations used, but many organisms can do without it under natural conditions. These issues of information-related strategies will be discussed in more detail in the next chapter.

Informational domains

The final piece to an account of natural information that characterises it as both specifying *and* including a local and probabilistic element will be to spell out the relation between the local and the probabilistic element in more detail. When moving from information$_L$ into her softer but supposedly often more relevant information$_C$, Millikan appears to target two related ways for that information to become 'soft', which I outlined in the previous section as NI-2b and NI-2c: first, the regularity *I* of the signalling relation might turn out to be statistical rather than nomological in kind; second, there might be regions in space and time – christened "domains" by Millikan – to which that regularity is restricted. The notion of informational domains in particular will become important in an inquiry into informational environments in Chapter 6.

(NI-5) The *domain-boundedness* condition: informational relations may be bound to certain domains *D* of some set of signals $\{r_1, \ldots, r_n\}$ or types of signals $\{r_1, \ldots, r_n\}$, where

(a) *D* is a spatio-temporally circumscribed region in which type **r** signals are present and detectable that correlate with some individual *s* or type **s** with a given degree of probability $p = n$, where $0 < n \leq 1$, as distinguished from other regions where $p < n$ or $p = 0$, providing for local maxima and minima;

(b) the extension of *D* is demarcated by empirical regularities I_e, as they apply to individuals, species and other spatio-temporally restricted entities at *s*;

(c) type **r** signals may pertain to different domains $\{D_1, \ldots, D_m\}$ at once and in non-contradictory fashion, depending on the regularities described in NI-5b;

(d) there is a possibility for a receiver *R* to keep track of one or various types of signals within *D*, under conditions described in NI-4, where *R* additionally may but need not be able to *identify* the extension of *D*.

More often than not, as Millikan's argument in (2004) goes, the only regularities at hand for a consumer of natural information are of a statistical kind. For these regularities, in turn, spatio-temporally local maxima and minima can be detected in many cases. If there are such local maxima and minima, these may serve as indicators for informational relations that concern individuals, populations, biological species or other kinds of entities whose identity is determined historically or genealogically. The laws of nature will provide some boundary conditions that, however, taken by themselves, will be insufficient to determine the identity of individuals and species or their possibility of existence – which may turn out to be the one thing relevant to *R*. Or those laws will not be sufficient to determine the specific local character of a bundle of conditions – why some individual or species not only could but *had* to exist here and now.

Basically, both 'hard' and 'soft' natural information are relevant to perception and action, but in different respects, on different levels: information_L concerns basic invariants in the environment, and the conditions for the perception of objects in the ambient optic array (what basic substance characteristics, what size, how distant, how transformed in visual in appearance when the perceiver moves). These conditions are, except under the most extraordinary circumstances – which we are most likely to find in some philosopher's possible worlds – not specific to any domain. Information_L, for its basic, foundational character, does not change across domains but first enables the tracking of individuals or kinds of things throughout their specific domains. Without some core regularities, environments would be capricious and unpredictable. The validity of information_C, in turn, depends on both the concrete informational domains that the perceiving organism is accustomed to, and the relative reliability of the signals that make up these domains.

Hence, information_L and information_C will also differ with respect to their reference classes (see NI-2 in the previous section): information_L comprises all and only those detectable relations in the environment which are governed by strict, and universally applying, natural regularities. Information_C, however, comprises relations that depend on less strict and less universal regularities by their very nature, and hence have their reference classes constrained by where and how those regularities apply.

It might seem that domains, as defined previously (especially in NI-5a), are thus a special case of reference classes, as defined previously (especially in NI-2b), so that $D \subset C$. There is a difference though: a domain is a domain of a type of signals, bound together by what is being tracked, or supposed to be tracked, by *R* (an individual, a population of individuals, a set of entities) and the I_e-regularities that respectively apply. A reference class, in turn, comprises whatever set of proximal *r* and distal *s* entities and a regularity *I* between them, which together shall serve to determine a probability with which some *r* within that set signals some condition at *s*. The choice of the reference class is, like the identification of a domain, dependent on the purposes, constitution and abilities of *R*. Hence, the reference class of a certain type of informational relation might be partly coextensive with the domain of a type of signals (namely in comprising these signals), but it makes no reference to why it should be circumscribed in a certain way, whereas a domain of some type of signal is explicitly defined with respect to the circumscribing condition, namely the presence of some individual, population or set of entities to be tracked by *R* (see NI-5b). The domain of a type of signals is a natural way of carving out a reference class for the informational relations involved, and its spatio-temporal boundaries.

In order to illustrate the problem to be solved by introducing the concept of domains of natural information, a brief discussion of Millikan's own example in (2004, Chapter 3) which she, in turn, borrowed from Dretske (1988), will be helpful: if tracks of a certain shape in the woods of her (or mine, or someone else's) home state have hitherto always and exclusively been created by quail (or, to abstract from concrete species, "*q*-birds"), they carry information about *q*-birds with a probability of 1 and, one might believe, they do so with lawful causal determinacy, as no other animal in those woods would be able to create tracks of the same shape, size and relative placement. The domain of the signal will be the entire set of *q*-tracks in those specific woods, and the reference class will comprise those tracks, the *q*-birds and the regularities by which *q*-birds give rise to *q*-tracks, which will be quite strict under this perspective. Nothing, however, rules out the possibility that pheasants (or "*p*-birds") may happen to cause tracks of the very same appearance, save for the fact that there are no *p*-bird populations in that area. What information, then, would tracks of that shape carry if a population of *p*-birds happened to migrate, say, from Tyrol to Carniola? And what would warrant the lawfulness of the correlation between tracks of that specific shape, size and relative placement in the woods of Tyrol, save for the fact that *q*-birds have not happened to migrate the other way? Finally, even if there were no such species as the *p*-bird to exist at the time of observation in the first place, what would rule out the possibility for them to evolve at a later time, or to have roamed the same and other woods in bygone days but now having become extinct? Under this perspective, the domain of the signals involved might remain the same in terms of the set of *q*-tracks in a given wood under investigation, whereas the regularities and possible source conditions included will have changed. The reference class will now encompass *q*-birds and the patterns of *p*- and *q*-bird population dynamics.

If there are *q*-birds as populations with a limited extension in space and time, and if there is a possibility of *p*-birds existing at other times and in different places while leaving behind identically shaped tracks, the information carried by those tracks would end up being equivocal and fall prey to the problem of disjunctive content that, according to Jerry Fodor (1990), bodes ill for an evolutionary naturalism about content. Thus, on Dretske's definition, it would not be information at all (although he allows for a softer notion of information in other places).[3] Millikan's alternative is to characterise natural information$_C$ as probabilistic *and* locally bound – which, as I have begun to argue in the previous section, are not precisely the same thing.

On this background, domains of natural information or "natural signs" are to be understood in accordance with the mathematical meaning of the term "domain" in the first place. The mathematical meaning is that of the domain of a function, that is the set of input values for which the function is defined (e.g. all real numbers or all positive integers), and from which the output values, as the "image" of the function, are mapped onto its target set (Millikan 2007). When transferring this notion to the realm of natural information, a domain is described by the mapping rule by which a series of signals $\{r^i, \ldots, r^n\}$ relates to a source or a set of sources $\{s_1, \ldots, s_n\}$, where that mapping rule is determined by causal, genealogical or other regularities I_e. These regularities become part of the explanation of the recurrence or reproduction of *r*-signals on occasions of *s* being *F*, and they also play the role of a necessary condition in explaining the presence and persistence of the mechanisms in *R* for detecting *r*-signals.

A certain ambiguity is introduced into this mathematical analogy when Millikan says that the domains of natural information comprise not all possible values that could be generated by some function (which, in some cases, may be indefinite to begin with) but those which have, in fact, been realised at the time t_n of *R* recording the signal. Under this interpretation, what is called the domain would actually be the *image* of the function, that is a region in space and time in which the mapping relation in question has been instantiated for a finite set of input values, and where it did so in accordance with that mapping rule which defines the domain of the function in the former, properly mathematical sense. A mere series of concrete instantiations of a seeming regularity by itself would be insufficient to warrant that regularity, because some of those instantiations might have different aetiologies. In the light of this observation, the most consistent practice would be to refer to domains in the proper mathematical sense, where it denotes the set of 'permitted' input values for a function defined for some mapping rule. But then, if we are talking about individuals and historical entities, such as populations and species, the history of concrete instantiations will be a constituent of the empirical regularities that make up some population or species. After all, the domains of natural information related to higher mammals would be differently shaped, in terms of available variation, environmental conditions and selective pressures for their ancestors, if the dinosaurs had not become extinct. Higher mammals might not even have evolved in the first place.

There is a recursiveness to the definition of domains of natural information in all cases that rely on regularities that are, partly or wholly, historically founded. Conditions that have obtained at some time and place will enable or constrain regularities at some later time and place. If we treated domains as the set of concrete instantiations of some informational relation and thus as the image of the mapping functions involved, that recursiveness might easily turn into circularity: the domain of a natural sign would then be whatever happened to have co-occurred to date, and the mapping rules could be identified only by recourse to that history. Strictly speaking, they would not even be rules but, with some likeliness, amount to mere enumerations of instantiations of co-occurrences that may be or may be not properly supported by, and would be entirely uninformative about, underlying regularities at the source.

On the view defended here, and certainly on the view that Millikan defends, too, informational domains are more than arrays of spatio-temporally proximate and/or phenomenologically similar but otherwise discontinuous, "gerrymandered" or randomly compiled sets of world affairs in which some seeming correlation can be detected by *R*. The definition and extension of domains would then be fully dependent on *R*'s perspective, with conditions at the source having at most a limited bearing on the nature of the informational relations utilised by *R*. The limitations of *R*'s perspective, given that *R* is a being of a specific constitution, with specific abilities but limited resources and a limited existence in space and time, do have an effect on what domains of natural information are used by him. On this level of analysis, where purposeful creation of information is not considered, the domains as such are constituted by regularities that may involve the nature and behaviours of the receiver, but they are not constituted *by* the receiver. That one signal may belong to different domains, where those domains may be found to carve up the world in different ways and to different uses, owes to the applicability of different but non-contradicting mapping rules to the otherwise observer-independent regularities involved (see NI-5c). For example one rule may circumscribe the domain of an individual, whereas another defines the domain of the population to which that individual belongs, so that the latter domain comprises at least some of the signals pertaining to the individual, too, and vice versa.

Hence moving from mathematical analogy to types of concrete cases, domains of natural information might be constituted in one of three fundamental ways:

First, informational domains may be results of *natural history*. A shadow of a certain shape circling overhead in a particular way conveys information about the presence of an Andean Condor in the Andes and of a California Condor in the mountains of the American West, and is unlikely to convey information on the presence of any species of Condor on the rare occasion of occurring anywhere in the Alps. Events of speciation, population dynamics, migration and extinction define whether, where and when a signal of a certain shape matches conditions at a certain source. The shape itself and its causal relation to the source, taken by themselves, are insufficient to achieve unequivocality in these respects. As in the *p*- vs. *q*-bird case, an indistinguishable shape could be created by a phenotypically and behaviourally very similar but genealogically rather remote bird. In order to make use of that information, the consumer would have to stand in a certain relation to those birds. However, to the extent that California and Andean Condors are phenotypically and behaviourally very similar to each other while inhabiting spatially discontinuous but ecologically similar regions, their competitors in the respective regions are not in need, and normally do not have the means, of distinguishing between the two species. A competing scavenger who would be moved by the appearance of a such-shaped and thus-moving shadow to defend his spoils against the Condor (or, more frequently, abandon them, as Condors are the dominant scavenger in the food chain) or take his presence as a signal of the presence of carrion nearby would be rather indifferent to the distinction between California and Andean Condors. The distinction is more relevant to a field biologist or conservationist surveying Condor populations. The Condor's competitor would not be indifferent, however, to the distinction between a Condor and a behaviourally very different disposed bird, say a predator of minor scavengers like him, mimicking a Condor's appearance in the sky with deceptive intent.

Second, domains may be composed of individuals or *aggregates of individuals*, for example if we try to follow one specific Condor specimen or population through space and time. Not just any such-shaped and thus-moving shadow within either California or Andean Condor habitats would do. The observer will have to use other information within the domain of the specimen, but transmitted through different, independent channels, to identify the shadow as a signal of the presence of *this* individual bird. Through various channels, the observer has to remain focused on, and keep track of, the domain of this individual in order to properly retrieve and use the information emanating from it. California Condors, for example for being extremely rare and critically endangered, are subject to a population recovery programme which includes the marking of individuals with numbered tags and transmitters – which makes for much easier tracking than having to rely on information collected in the wild. Aggregates of individuals need not be phylogenetically related though. Depending on the receiver's needs and purposes, they even could be simply that in some cases: aggregates of individuals in specific constellations and at certain times and places that are relevant to him. In such cases and only in such cases, the regularity governing the domain is partly constituted by the receiver's constitution and needs.

Third, domains may owe to universally *law-like* rather than locally bound probabilistic regularities on a physical level but are circumscribed by *concrete conditions of realisation* that only hold locally. For example a hygrometer will not convey any information on atmospheric water content when placed on Venus, nor will there be any use for a barometer or olfactory organs in interstellar space. The conditions on Venus or in interstellar space do not fall within the range of values for which the function or functions of the respective detectors are defined. Under these conditions, although the operation of the instruments and organs can be expected to accord to universal laws of nature, they will not be able to execute their functions, namely the collecting of natural information. This is not to say that they do not have any function or lose their function when hence placed outside their range of operation. The informational relation they are supposed to detect is not present where they are, and/or the channels they would need for detect that information are missing or misaligned, all for perfectly law-governed reasons. Some receivers of natural information might be able to discern between those regions in space-time where the conditions are in place for the presence and/or transmission of the information in question and those regions in which they are not. Or the receiver might be physically constrained to remaining within the region or merely happen never to leave the region in which it holds.

Dretske's famous magnetotactic bacteria may serve as a paradigm of lawfully bounded domains of the latter kind (Dretske 1986, 26–28): For those anaerobic marine bacteria, about the only relevant variable in their environment is the oxygen content in the waters they inhabit, which is strictly, and rather lawfully, correlated with depth below the water surface. With increasing depth, the concentration of oxygen they will encounter reduces. Magnetotactic bacteria, however, do not have any sensors for oxygen content in their surroundings. They do not perceive it; nor is even the most simple of stimulus-response mechanisms involved. What they do is to track a correlate of that gradient that also holds with law-like regularity, namely the direction of the Earth's magnetic field. The lines of the Earth's magnetic field are aligned, as for any other magnetic dipole, as field lines, which can be described by any smaller magnetic particle being placed in the field. These lines emerge from the magnetic south pole, bend outwards towards the equator and return to Earth at the north pole. As a result, the lines will point upward relative to the surface of the Earth in the south and downward in the north. The magnetic particles enclosed in the bacteria's cell membrane align with the field line. In this sense, magnetic orientation is a lawful proximal correlate of the vertical gradient of oxygen concentration. This is where the bacteria will invariably move. However, this direction is inverted from one hemisphere to the other, so "northward" will be "upward" in the south and "downward" in the north. Each specimen of the respective subspecies of bacteria must be placed on the right hemisphere in relation to its northward- vs. southward-oriented constitution in order to maintain the mapping between field direction and the gradient of oxygen concentration on which it depends. With no supplementary sensory facilities to warn it, the bacterium cannot identify the domain of the downward signal, however lawful the correlation.

The first and the second of the ways of carving out domains of natural information are the ones Millikan has in mind, whereas the third seems to be implicitly subsumed under the former two or, more probably, under their preconditions.[4] The first two ways even fall into one if one conceives of species as historical individuals (as in Millikan 2000) rather than extensionally defined populations. The choice of species concept makes a difference to the interpretation of the nature of the informational relations involved. Quite obviously, if a species is to be treated as an individual, and if its being an individual is not considered a theoretical fiction in the service of explaining biological phenomena, but a real entity held together by a well-defined set of genealogical relations of descent and proximity, the observer's task, when attempting to track the domain of the species, lies in tracking the domains of the individuals belonging to that species. The domain of the species specifies the domains of its individuals. Misidentifying some r as a signal of a q-bird qua q-bird rather than merely of a such-shaped (p- *or* q-) bird leaving such-shaped traces, amounts to a failure at correctly recording natural information related to that individual *as a member of the species*, although it may still be correct of that individual *as an individual* (that bird in my garden last night, or that bird gracefully circling overhead in the Andean skies).

If, in contrast, species are extensionally defined, the definition of their domains is observer-dependent in a very specific way: Signals $\{r^i, \ldots, r^n\}$ emanating from different individuals $\{s^i, \ldots, s^n\}$, where the mapping of r^i onto s^i, r^j onto s^j and so forth is supposed to be unequivocal on the source side, have to be subsumed by R under a common type of signals **r** that correlates with a type of individuals **s**. That mapping is not, or not fully, determined by the natural information carried by the individuals $\{s^i, \ldots, s^n\}$. The signals might be marks of type-identity in properties for the individuals, but there will also be signals of properties in which these individuals differ. Hence, R is endowed with the task of identifying the similarities and differences relevant to his purposes without the benefit of an additional property that would help him to type $\{s^i, \ldots, s^n\}$ together in an unequivocal way and thus warrant *R*'s acts of identification. Any misidentification of an individual will be a misidentification only in relation to the practical purposes of *R* and only if these practical purposes are adversely affected by that misidentification. The practical purposes of the competitors of California Condors will not be affected when an Andean Condor accidentally appears in Californian skies, given their phenotypical and behavioural similarities, whereas the practical purposes of a California Condor looking for a mate and the practical purposes of the California Condor Recovery Programme are likely to be affected.

If we restrict our view to *practical* purposes, the latter observations will equally hold for either species concept. Millikan (2004, Chapter 18) applies the term "practical kinds" when referring to those situations in which a difference between species or other types of things or even between one and the same vs. different individuals is practically irrelevant to *R* since the effects on *R*'s purposes and behaviours of treating them as identical are negligible. Thus, it will also be irrelevant in practical terms whether the species involved is a historical individual or an aggregate of individuals, as long as *R* carves them up in proper alignment with his practical purposes. The California and the Andean Condor form one "practical kind" for the competing scavenger, but two different practical kinds for other Condors, and two natural kinds for the field biologist and the conservationist. Even the distinction between repeated encounters with one and the same individual and with other members of the same species may not be important under all circumstances (see Millikan 2004, 219f).

However, the theoretical implications of the choice of species concepts may be notable, and some higher-level practical considerations may flow from it. Choosing species, understood as historical individuals, as the paradigm of domain-bound natural information might serve to mix up the local and the probabilistic character of natural information$_C$. Assuming species to be individuals implicitly suggests that the character of natural information$_C$ amounts to being probabilistic *qua* locally bound (see p. 65 in this chapter, see also NI-2b and 2c). This is no minor point, and it takes some explaining.

The extension of a domain of a species as a historical individual would be marked by its incipience at t_0, its population dynamics, its patterns of migration and its extinction at t_n. Prima facie, the sources of all possible equivocation in the definition of a domain of a species would hence lie with, first, determining its time of incipience, second, the lines of descent that flow from it, third, its population distribution and fourth, its time of extinction (which might be the easiest to identify if it has happened already). These are all things that *R* could be mistaken about, but not the species itself, as it were. The rule that determines when a species is a species is supposed to be unequivocal and observer independent, and hence the membership conditions of individuals pertaining to that species are supposed to be unequivocal and observer independent, too.

On this view, it should be a straightforward matter to account for the possibility that one may encounter, on the one hand, individuals that belong to the species in question but, being imperfect copies of their type, display some divergent characteristics and that, on the other hand, one may encounter individuals that display the seeming characteristics of the species in question but do not belong to it genealogically. Instances of the latter possibility would be the equivocal traces of *p*- and *q*-birds and the moving shadows of the 'wrong' species of Condor, instances of the former would include any organism with dysfunctional organs or other phenotypical aberrations. Prima facie, the same conditions will apply to types of artefacts: an individual Ford Model T will be a Ford Model T because it was designed by Ford Motor Co., and moved off the assembly lines at Ford Motor Co. factories between 1908 and 1927. A bootlegged car or an artful replica of the historical model, meticulously reverse engineered from an original Ford Model T, would not count, and *R* would be fooled if having it sold to him as a classic car from a fraudulent vendor. Conversely, a somewhat wayward Ford Model T specimen, for having been produced at Ford Motor Co. factories according to the Ford Model T design, albeit with some obvious aberrations, will still belong to the same type.[5]

However, just as pirated copies of digital media, every bit indistinguishable from the original and produced in the same way as legitimate members of the type but only illicitly so, partly undermine the seemingly clear-cut relations of descent for types of artefacts, treating species as individuals is exposed to some ambiguities contained in their very definition. For example it will be difficult to precisely determine a speciation event, both in terms of pinning down the moment when a variety becomes a new species (a problem that has haunted evolutionary theory ever since Darwin) and in terms of demarcating species from mere varieties in the first place (a problem that has been haunting natural history even before Darwin).

In these and many other kinds of cases that concern the question of membership of some *s* in a type **s**, if this membership shall hold by virtue of certain type-defining properties rather than definitional stipulation on *R*'s side, the rules governing the mapping between the domain of signals of an individual and the target set might turn out to be underdetermined. There may be boundary cases where the mapping of signals pertaining to a putative member of the type in question onto the target set, that is its being a proper member of the type, may be irreducibly equivocal. If this is so, some of the ambiguity that makes natural information$_C$ probabilistic may well reside at the *source* of that information. In such cases, the *local* character of this kind of natural information does not fall into one with its *probabilistic* character. The species would be unequivocally defined at the centre but less so at the spatio-temporal margins of its extension. Whereas the adherent of a population view will leave the disambiguation of conditions at the margin to *R* and her explanatory purposes, and find this unproblematic, the species-realist will be more likely to ascribe to *R* the responsibility for that very residual ambiguity and to remain (implicitly) committed to a fundamentally non-probabilistic view of natural information. He will also find it difficult to acknowledge that ambiguity as a matter of fact.

In most relevant situations, however, the practical purposes of the perceiving organism will dictate the rules of disambiguation so that he remains able to carve up the things he encounters into practical kinds that are suitable to the fulfilment of his purposes (see NI-4b). Residual ambiguities might assume some importance if and when conditions in the environment become unstable or change in ways that the receiver cannot easily anticipate and accommodate into the practically determined domains of the signals that he relies on. In such situations, the conditions of stability (NI-4c) and unequivocality (NI-4d) of natural information have been violated and call for some kind of adaptation or, less biologistically speaking, readjustment.

Resurrection at last

If I have to summarise the discussion in the preceding chapters in two brief points that pick up on the key authors on which I built my account, it will be these: first, despite acknowledging the irreducibility of situations of underspecification for a perceiving organism, both Dretske and, albeit less unequivocally, Gibson, establish natural information as a specification relation. Second, despite his reference to knowledge, the Dretskean notion of information from which my inquiry commenced partly parallels Gibson's deflationary spirit with respect to knowledge and mental representation – which is ultimately the more coherent view.

It is these two parallels that I see as deserving to be defended or, as the title of Chapter 2 had it, "resurrected". At the very beginning of *Knowledge and the Flow of Information*, Dretske says: "In the beginning there was information. The word came later" (1981, vii). He thus anticipates that the historically and systematically prior thing for organisms is to "selectively exploit" information in the environment to adaptive ends, that is the guidance of activities. Dretske correctly identifies Gibsonian information as "higher-order invariants in a temporal series of signals", but then continues to refer to sensory representation (Dretske 1981, 145) – a notion that Gibson would not have subscribed to. In the endnote to that passage, he also mentions that a distinction has been made between Gibsonian information for perception and information under the mathematical paradigm, including by Gibson himself (Dretske 1981, 255, n10). He then goes on to quote a passage from *The Senses Considered as Perceptual Systems* in which Gibson says that "a property of the stimulus is univocally related to a property of the object by virtue of physical laws" (Gibson 1966, 187). It seems, however, that Dretske quotes this passage in order to actually *relativise* the very distinction between Gibsonian and mathematical information – which should not be surprising because Dretske himself apparently believes his account of information to be in line with both Gibson's and the mathematical paradigm.

In this attempt to align himself with two concepts of information that are demonstrably disparate, one can detect, first, the tension between Dretske's deflationary view of information and his epistemological aims and, second, the subtle but significant difference between Dretske's and Gibson's accounts: on Dretske's view, informational relations are invariant relations between world affairs that could, in principle, be detected by a receiver, whatever his constitution and abilities might be. These relations can be analysed in isolation from each other, and may be integrated by a suitably disposed observer, who is viewed in detachment from his environment, in inferential processes of knowledge acquisition. On Gibson's view, informational relations are invariants in the patterns in the ambient energies that can, in practice, be detected by a perceiver on the concrete grounds of his constitution and abilities. Some of these invariants, as I will further elaborate in the following chapter, form specific integrated bundles for the organism that are determined in relation to what he can and needs to do in his environment. Assessing those invariants in isolation would not do justice to the complexity and multi-modality of interactions. By virtue of being thus integrated and situated, the information available in the environment will paradigmatically be sufficient to directly guide the perceiving organism towards the intended state of affairs, with no *k*-condition needing to intervene.

If we want to know how animals and humans, to return to the quote that opened my discussion of Gibson (see Chapter 3), have come to "communicate with cries, gestures, speech, pictures, writing, and television" (Gibson 1979, 242), and how they have thus come to be knowledgeable creatures, we might wish not to include the knowledge criterion from the start, as most organisms do quite well without it most of the time. One might also come to more precisely identify the locus of informational content, namely in the specific relation among, first, what information is present in the environment, second, what of this information is available to the perceiving organism and third, to what ends he uses it.

Notes

1 The possibility of equivocal symptoms is mobilised against Dretske's nomological account of information by Millikan (2001).
2 The different answers that can be given to this question are analogous to the – somewhat notorious – distinction between *K*- and *r*-selection.
3 One example for a more liberal approach in Dretske would be the very passage in Dretske (1988, 56) in which he introduces the quail vs. pheasants example, as Millikan (2004, 32) notes.
4 In Millikan (2004, 52), she says: "Don't trust what looks like that needle when you go up to Mars, but here on earth, the positions of needles on gas gauges all fall in the same roughly defined locally recurrent sign domain." By implication, gas gauges belong to a type of artefact that shares a number of relevant properties with species, namely being reproductively established as a type of artefact, which is distinguished by its proper functions, and which populates only a limited set of slices of space-time. Here on earth, we recognise what a gas gauge is and what it signals when we see one. Similarly, don't trust your olfactory organs when you go to the moon and seek to identify objects on the moon by smell – but we know the shape and function of olfactory organs of different species, as established in processes of biological evolution, and we know something about the conditions for performing that function.
5 This exemplary juxtaposition characterises one of the core tenets of Millikan's theory of proper functions, as proposed in Millikan (1984) and discussed in more detail in Chapter 7, namely that of (first-order) reproductively established families.

5 Making an environment

We should now be equipped with a concept of natural information that is more detailed and specific than James Jerome Gibson's negative claim that information is not best understood on the model of communication, and his positive claim, placed right next to the former, that information "refers to the specification of the observer's environment" (1979, 242). Although these claims are closely related, it might be helpful to consider their bearing on my core argument separately.

Taking up Gibson's negative claim, we can now ask for the implications of the theoretical notion of natural information developed in the previous chapter for non-natural information, hence the very kind of information that Gibson characterised as a misguiding model. To what extent is information that has been purposefully produced and added to our environments to be understood in the same way as natural information? I will reserve most of the discussion this question for Part II but start working towards it near the end of this chapter and in the next.

Taking up Gibson's positive claim, we can now ask for the role of the environment proper in both being *specified for* the organism by the information therein and being *specific to* the organism – and in often leaving him with situations of underspecification. Even though natural information is an objective commodity, human and other natural beings may get a variety of things wrong in a variety of ways, without necessarily being exposed to the charge of error and dysfunction. However, they will also encounter opportunities for adapting their environments and the information therein to their needs, so as to make them more specific both *for* and *to* themselves.

Given that the creation and use of artefacts is a special case of making the environment more specific, I will discuss the latter point first, and dedicate this and most of Chapter 6 to it, with respective focus on the shaping of environments and the nature of informational environments. I will begin my discussion with explicating the requirement, mentioned only in passing earlier (see p. 54 in Chapter 3), that both the historical and the ecological context of perception and behaviour have to be considered. These contexts are intertwined to the extent that a natural history of every organism and of every trait of an organism is identical with a history of the environmental conditions under which it emerged and by which it was shaped, whereas any environment is a result of history of particular interactions between populations of organisms and the physical and biotic conditions under which they acted and interacted. These interactions, I will continue to argue, include various modes of modification or "construction" of conditions in the environments by the organisms themselves.

History, ecology, environment

Information specifies the environment for the observer but that specification, as I have sought to demonstrate, always obtains with respect to the constitution and abilities of the organism perceiving and acting within the environment that he inhabits. *Why* the organism's constitution and abilities are shaped the way they are is the question of *historical context*, which specifies both the selective pressures to which an organism or a population has been exposed over time, and hence the processes of adaptation, and the adaptive functions that result from the selection of those variant forms which have been produced over time. The selective pressures to which an organism or a population are exposed are determined by no fewer than two factors in conjunction: first, the traits already present in the organism or population at t_0, and second, all relevant conditions encountered in the environment, that is the status and the rate and intensity of changes of those variables which affect the organism from t_0 onward.

What is not implied here is the assumption that all traits of an organism are to be viewed as adaptations and that all adaptive traits are a direct and exclusive product of natural selection. The possibilities are preserved of the existence and relevance of "spandrels" and "exaptations". Whereas spandrels are adaptively neutral, hence function-less, traits that arise as secondary effects of adaptive ones or are shaped by constraints on form (Gould and Lewontin 1979), exaptations cater for the possibility of adopting existing traits for new functions (Gould and Vrba 1982).[1] Nonetheless, presuming that complex abilities are ultimately to be understood as adaptive traits, and that natural selection will have made an indispensable contribution to shaping them, the Neo-Darwinian view appears defensible. Even so, the further implication shall be avoided that organisms were passively exposed to selective pressures that work in analogy to external forces on inert matter. After all, the organism-environment relation is of a two-way kind, both in terms of an organism's abilities to detect and respond to conditions therein and, more basically, in conditions environment- and organism-bound conditions affecting each other.

That latter set of relations defines the *ecological context* for the organism, and thus *how* the organism is shaped in his specific constitution and with his specific abilities, so as to respond to conditions in the environment that he inhabits. An environment in the sense to be discussed here is marked by two important properties: it is to be considered dynamic on different levels of description (discussed further in the last section of this chapter), and it is, in a qualified sense, specific to an organism. Any type of ecological reasoning, from plainly biological to normatively political (the latter of which will not be considered here), will share at least these two broad characterisations of environments as *dynamic* and *specific*, although different approaches are unlikely to agree on all details.

As to the property of specificity, the notion of an environment, if it is to be useful in biological ecology, should neither be plainly equated with an organism's or other system's physical surroundings, nor should it be confounded with the subjectively experienced environments of the organisms inhabiting a certain region in space-time. A well-known attempt at providing an intermediate account was undertaken by Jakob von Uexküll (1956). On his view, there would be a plurality of differently constituted individuals living in a plurality of differently constituted environments, each pair being intertwined in their specific fashion but along similar patterns of interaction (a "Funktionskreis" or functional cycle). The perceived, phenomenal environment will be at variance between different organisms, as will the relevant variables within the environment and the possible ways for the organisms of influencing conditions in their environment, in scope and depth. The environment would even be different for different members of one and the same species, for the rather trivial facts that they inhabit different slices of space-time and that their constitution, perspectives and behavioural opportunities will vary, however slightly. What remains identical across all those individually constituted environments is the general functional pattern of interaction. However, this usage would make it difficult to account for how those many individual environments are related to each other, and on what grounds.

Conceiving of environments in the Uexküllean way is instructive when exclusively focusing on an organism's phenomenal environment, but not when seeking to account for how different organisms interact in shared surroundings. Moreover, his view will be fully tenable only within the neo-vitalistic framework within which it was developed. The teleological structures within which an organism is embedded could then be taken for granted rather than being the explananda for a theory that seeks to illuminate the ways in which some such structure emerges and is maintained. In absence of the neo-vitalistic premises of von Uexküll, and in view of accounting for the interaction of various organisms in a shared environment, a promising solution will be to conceive of environments in a way that echoes the intermediate, twofold nature of Gibsonian affordances, as being specific to an organism but objectively rooted.

A conceptual distinction between different notions of an organism's environment that highlights its twofold nature has been introduced by Robert N. Brandon (1995), within the context of theories of natural selection. It was crafted in order to distinguish between those instances of differential reproduction within populations which are and those which are not due to the environmental fit, hence the adaptedness, of certain traits of the organisms involved rather than sheer luck. Under "external environments", Brandon subsumes all physical and biotic conditions in an organism's or a population's surroundings that may affect them and the local or temporal variance in these conditions. These conditions are typically directly measurable by an external observer. The "ecological environment", in turn, is constituted by those factors within the former set that have an effect on, and thus are relevant to, individual organisms with respect to their rate of reproduction, and hence their contribution to their population's demographic make-up. Precisely to the extent that the organisms within that population are genetically homogeneous, variant rates of reproduction implicitly measure the variance in environmental factors. Finally, the "selective environment" is made up of those factors which account for differential rates of reproduction between genetically heterogeneous subsets of some population. Here, the distribution of phenotypical traits among the population, if and when affecting rates of reproduction, becomes part of the environment, too.

As an illustration for the distinction between ecological and selective environments, Brandon (1995, 47–49) uses the image of a field with soil conditions that vary over certain areas or that change over time. If we plant seeds from a genetically homogeneous stock in a regular pattern, the patterns of growth and health of the plants that emerge will provide a measure of variation in the soil conditions that have an effect on the plants, allowing us to identify a number of relevant variables within the plant species' ecological environment. If we plant seeds from two or more different varieties in the same manner, the variant patterns of growth will help us to identify a number of relevant variables within the different varieties' selective environments. For example, one variety might be more affected by a change in soil conditions than the other, so the variant patterns of growth and reproduction will provide a measure of the relative fitness of the respective varieties. In this sense, the organisms themselves can be interpreted as measuring instruments that are able to determine the conditions in these two latter types of environment.

Hence, the line of distinction between the three levels of description of an environment is drawn by distinguishing levels of conditions outside and within a population with respect to their objective relevance to that population and its individual members. Although the external environment can be defined in disregard of the constitution, the abilities and the needs of the particular organisms inhabiting it – apart from the question whether the reference point for "external" is the individual or the population – the latter two can be defined only in relation to the properties of individuals and populations. Nonetheless, there is, strictly speaking, no ontological difference among external, ecological and selective environments. There are only different levels of description that refer to different relations between the constituents of an environment, and their respective relevance to a population and its members.

The point that Brandon seeks to make with his threefold distinction is that external, ecological and selective homogeneity vs. heterogeneity of some environment may not necessarily reflect each other (Brandon 1995, 64–66): a homogeneous external environment may be heterogeneous in ecological terms, as the key factor of reproductive relevance may lie with other members of the same population. Such would be the case for a high population density of adult specimen in some area that gets in the way of the growth and reproductive chances of juveniles. Even developmental stages of an organism may become selectively relevant. Conversely, although there will be no selective heterogeneity without ecological heterogeneity, a physically and ecologically heterogeneous environment may still be selectively homogeneous, as relative fitness of genetically variant sections of the population may covary with variation in the ecological conditions, that is the variant forms may be identically affected by the ecological conditions in question.

Although the focus of the Brandonean analysis quite naturally lies on the selective environment with respect to genetically variant populations, the ecological environment and the homogeneity vs. heterogeneity therein with respect to various individuals will be of main concern here. The general upshot of the Brandonean distinction is that different sets or bundles of – objectively given – conditions obtaining within the same spatio-temporal surroundings will matter in different ways to different organisms, or even to the same organism at different times, and that these conditions are measurable in principle. In the context of ecological considerations, this means that patterns of similarity and difference among a variety of conditions, that is patterns of distributions of values across different ranges of, partly interacting, variables will account for different *types* of ecological environments. We will encounter savannahs, rainforests and even urban environments in a variety of different geographical regions. As such, these environments are not characterised by spatial proximity but by the presence of certain values and certain degrees of variability, within variant ranges and with variant frequency, for a number of variables. To these conditions belong temperature ranges, the presence or absence of seasonal changes or patterns of precipitation vs. evaporation but also the presence and behaviours of other organisms. The values of some variable or combination of variables will have to remain within a certain range for some organism in order for him to persist and reproduce. He will either have to pick up information about those values and their transformations or he will have to be reliably placed within an area where they are, and remain, suitable to him.

If it is the values and ranges of variability of some environmental variables that define an ecological environment for an organism, these environments may but need not coincide with a certain physical location or contiguous region. For example if the values for a number of environmental variables show only minor variation within the range of habitation of some organism, and if the organism depends on relative constancy of those values, he may not ever be challenged to react to any major changes, in rate and intensity, of the respective values. Nor will it be important for him to collect information about changes or imminent changes in those not-so-variable variables, as long as he remains confined to living in surroundings of low variability. In the tropics, much unlike most of North America or the Central Asian steppes, temperature will be one such low-variability variable. There will be some temperature-related information in the organism's surroundings, but neither is there much of a reduction of possibilities at the source in play, as the range of values to be assumed is limited to begin with, nor is the extension of the domain of those signals actually *small* enough to matter. If the domain of those signals, in spatio-temporal terms, is coextensive with or larger than the organism's range of habitation, and hence if the organism is unlikely to ever leave that domain, he is equally unlikely to encounter conditions in which he would have to keep track of that domain. He may be physically unable to climb nearby mountains with lower and more variable temperatures (he might be a plant, after all). Alternatively, he may be informationally bound to his tropical lowland by a reliable proxy of the temperature gradient, such as the disappearance of his main food source with increasing altitude and decreasing mean temperature.

In either case, relatively constant temperatures are part of the background conditions for the respective organism. These are conditions on which the organism relies, and which will have a direct or indirect bearing on the fulfilment of the organism's goals, but which may not be directly perceived by, or even be directly perceivable to, the organism. Continuing with the present example, temperature constancy will not be part of the background conditions for a hardy plant that evolved in a more variable climate and is hence adapted to endure, and react to, major temperature swings. Nor will constant temperatures at some location belong to the background conditions for an animal that follows a regular pattern of variation in temperature and daylight over the course of the seasons across geographical regions. Such will be the case for a migratory bird, who visits a sequence of locations precisely in order to avail himself of suitable temperatures and a suitable supply of food. The conditions on which the bird depends have to be relatively constant, although locations will change according to whether they provide conditions that remain within the required range over some time. The bird is in the business of actively seeking out the conditions he requires. These are variant strategies of coping with patterns of variation in environmental conditions.

Adapting ecological niches

Both on the perceptual and the behavioural side, there is a dynamic aspect to environments, even if some environment is marked by a high degree of constancy or by stable recurring patterns for some relevant variables. Let us recall that one of Gibson's primary concerns is that perception is an activity in which the perceiving organism interacts with an object of interest over time, and that this activity is based on tracking its invariant aspects. In a similar vein, tracking the domain of a signal, as suggested by Ruth Millikan, means to keep up with the behaviours and transformations of conditions at the source that are mapped by a multitude of signals within that domain.

On the perceptual side, information related to some source is *not* collected in one-off encounters with static states of affairs. Typically, an organism will not come up against things that do not move, cannot be moved and look the same from every angle, from any distance and under all lighting conditions. Such static conditions for perception do not even apply to situations of looking at a picture – unless the viewing conditions are artificially restricted, as in what Gibson calls "aperture vision". Instead, a broad variety of signals related to a distal affair is collected over time, from different relative positions and under conditions where that affair is likely to change in various respects, in various ways. So much I consider to have established in the preceding chapters.

On the behavioural side, there is a case to be made for organisms not merely adapting to pre-existent conditions in their environments but interacting with, and thereby actively altering, states of affairs in those environments, some of which will be specifically relevant to them. An active adapting *of*, rather than merely *to*, conditions in the environment may occur on developmental, ecological and evolutionary time scales, with the possibility of overlaps being granted. Hence, it comprises effects that manifest themselves first, within the lifespan of the organism himself and his contemporaries, second, over the course of generations of a population and third, in the selectively relevant conditions for further evolution within a population. Moreover, the active adapting of conditions in the environment includes the creation both of recurring patterns, such as seasonal collecting and storing of food, and of more persistent effects or structures, such as built features of the environment or tools.

The concept of ecological niches in biology may in effect, if not in intention, turn out to obscure the dynamics and bi-directionality of organism-environment relations that are at issue here, and sometimes even their specificity. The earliest concept of an ecological niche, as introduced by Joseph Grinnell (1917), still referred to niches as the particular delimitation of a concrete species' habitat, and hence as tailored to that species, with the aim of describing the current status of the environmental relationships of an organism by a naturalist. For its focus on as-is conditions, a perspective on the dynamics of organism-environment relations was not part of this account, whereas specificity was the aim by definition.

In the meantime, a concept of ecological niches has become entrenched that compartmentalises environments into specifications of functional roles for adaptive traits and maps them onto types of spatio-temporal contexts. An ecological niche, on this influential view, which was proposed by Charles Elton (1927), actually *is* an organism's functional role, or "its place in the biotic environment, *its relation to food and enemies*" (Charles Elton 1927, 63–64, emphasis in original). Defined in the Eltonian way, there would be niches, for example for aerial predators specialised on terrestrial animals, with sub-specifications for prey size; that niche would require certain conditions regarding terrain and climate to be present, and it would be populated by eagles, hawks or falcons; it might have been populated by Pterosaurs during the Cretaceous; the niche for aerial scavengers would vary from the former in several respects, and would be populated by Andean or California Condors or any other organism who meets the specifications for the niche. Thus defined, an ecological niche might be occupied by various species at different times, in different places, and it might remain unpopulated in some places, at some time. An ecological niche is hence considered stable over place and time, and it is specific only with respect to functional role specifications. All the dynamics of changing environments would leave the functional roles and matching types of locations intact while allowing both for types of locations being realised in various places over time and for ecological equilibria to be upset, so that inhabitants of niches become dislodged and possibly replaced by new tenants, resulting in the establishment of new equilibria.

In another classic definition that competes with Elton's, Evelyn Hutchinson (1957) distinguishes between the "fundamental" and "realised niche" of a species. The former is defined in abstraction from all constraining factors, such as competition and predator-prey relationships, as a hypothetical "*n*-dimensional hypervolume", to be described as a geometric function, and composed of the coordinates of all values of all relevant variables in the ecological environment of some species. That space is bounded by the limit values of those variables that, in the absence of concrete constraining factors, would be compatible with the continued existence of the species under consideration, and it will thus be filled by the suitable ranges of these values. The realised niche is the sector of this hypothetical hypervolume that is circumscribed by a given set of concrete constraining factors. Realised niches, then, are always subject to the whims of the current conditions of realisation, and may change over time or may be changed by the organisms, for better or for worse. On the level of realised niches, a dynamic aspect enters the conception of ecological niches that allows for the notion of niches being modified by their inhabitants. On the level of fundamental niches, evolutionary change of the constitution of a species will be the exclusive provider of dynamics. In addition, specificity is acknowledged in terms of the variable dimensions and extension of the mathematical spaces of both fundamental and realised niches for each species. However, these notions are not systematically elaborated with respect to the possibility of organisms modifying the dimensions and extension of that space.

The Grinnellian, Eltonian and Hutchinsonian niche concepts are, first, partly idealising in an unfavourable way in that they abstract from concrete conditions of realisation and the specific properties and activities of specific organisms. In the case of Hutchinsonian niches, this critique applies to the notion of the fundamental niche, whereas the Grinnellian niche concept is altogether exempt from it. Second, these definitions are static to the extent that they do not sufficiently account for the possible effects of the behaviour of organisms on the shape of a niche, with partial exemption for Hutchinson's realised niches.

It is these two points on which Richard Lewontin's critique of what he deems the received view of organism-environment relations, and hence ecological niches, picks up. He argues that an environment is not merely a set of pre-existing conditions to which organisms would adapt in reactive fashion, by the trial-and-error process of random variation and natural selection. If this were so, those external conditions would be the fixed problems to which their adaptations were the fixed solutions (Lewontin 1982, 162). Conversely, an organism's traits, on the view criticised by Lewontin, would "map the demands of the environment through adaptation" (2000, 47). Instead, an environment, on the Lewontian view broadly shared here, is both dynamic in bi-directional fashion and specific to an organism:

> The environment of an organism is the penumbra of external conditions that are relevant to it because it has effective interactions with those aspects of the outer world.
>
> (Lewontin 2000, 48f)

To illustrate this point, Lewontin notes that one would find it hard to describe the environment of an organism without first describing the activities of that organism (Lewontin 2000, 50f). In his earlier essay on the same topic, he goes one step further when he says: "Indeed, an environment is nature organized by an organism" (Lewontin 1982, 160) – a view that may be aptly labelled as "environmental constructivism". Environmental construction, on Lewontin's account, will include any intervention by an organism or population into processes in the external environment that changes conditions therein in such a way that the intervention in question becomes a necessary part of the explanation of the nature of the organism or population and their adaptive success – or failure. The construction of environments is not to be understood in a literal sense, although the building of material structures might be involved.

On the face of it, construction, Lewontin-style, refers to the observation that some properties of the environment might be modified by behaviours of the organism, not only in terms of animals building structures in their environments, but more broadly in terms of certain variables being affected by the behaviour in question, where these variables might be relevant to the organism himself or to other organisms inhabiting the same slice of space-time, or a subsequent one. For example a herd of grazing animals is likely to affect the shape of the flora and soil conditions of their pastures, both in the short and in the long term. The effects of their grazing might also improve the conditions for reproduction of some inedible plants that would otherwise be outcompeted by those now eaten by the animals.

Lewontian environmental construction also includes cases where features of the environment remain physically unaltered while being treated in such a way as to affect the attainment of an organism's or population's goals. This kind of construction seems to be implied in his example of subspecies of *Drosophila pseudoobscura* and their humidity preferences (Lewontin 2000, 53f): one would expect that the subspecies from drier regions have a lower humidity preference than those from more humid regions, but the reverse was demonstrated to hold. The arid-environment subspecies live in small crevices in which the little moisture there is in those environments accumulates. In this case, the term "construction" refers to the activity of animals seeking out microhabitats with suitable conditions, which will count as their environments proper, rather than an activity of modifying conditions in that microhabitat.

These cases should be distinguished from a third type of intervention by organisms into environmental conditions that was also subsumed under the "construction" label by Lewontin: situations where the effects of an organism's behaviours undermine rather than foster the attainment of his goals or his chances of reproduction. For example the behaviours of a grazing herd might affect rates of soil erosion in their territory, thus possibly undermining their own chances of reproduction while, at the same time, being beneficial to algae or bacteria living off the nutrients swept downstream during this year's rainy season. Hence, the overgrazing herd might contribute to constructing the niche, and to improving the ecological environment, for some plant in some remote place while ultimately destroying its own ecological environment. Such accidental side effects only indirectly, and clearly only negatively, relate to the possible adaptive functions of the traits that produce those effects. It is here the latest where Lewontin's notion of environmental construction becomes too broad to be entirely useful, as Peter Godfrey-Smith (1996a) observes.

Theories of niche construction, having found their paradigmatic formulations in F. John Odling-Smee et al. (1996; 2003), with Meredith West and Andrew

King (1987) as relevant predecessors, may provide more focus to the notion of construction involved here. Broadly speaking, ecological niches are the specific sets of conditions in an environment that are reproductively relevant to a certain population. Different populations inhabiting one area at one time will depend on different sets of conditions at that place at that time. The make-up of these conditions partly depends on the activities of the organisms themselves. If there *is* some adaptive function to the construction of features of the environment, these features will become part of the necessary conditions in an explanation of an organism's success on a proximate behavioural or an ultimate selective level, or both. If these features are indeed part of the necessary conditions in question, there are good reasons to view them as coupled with the organism: he has to track the presence of some feature and put himself in the right relation to it, or he has to modify some feature in order to prevail. These necessary conditions, whether found or made, will make up his niche.

Although the views of niche and environmental constructivism are related, their terminology diverges, and with it, perhaps inadvertently, the general argument's focus. Ecological niches are specific by definition on the niche construction view, whether they are static or dynamic in nature. Hence, attention will be drawn to the latter aspect: are niches found or constructed? Environments, in turn, are specific only on some accounts – as, for example, for von Uexküll and, in some respects, Lewontin – but are more readily accepted to be dynamic affairs. Hence, the argument will gravitate towards the aspect of specificity: what particular contributions do organisms make to the dynamics of their environment? To be sure, these aspects interlock: if an organism makes a number of contributions to the shape and dynamics of his environment, some of these contributions will be relevant to himself, and hence be specific. If, in turn, an organism creates the niche that he inhabits, he will also have altered his environment. However, as we have seen, not all of the contributions that an organism makes to his environment will actually be constructive with respect to creating his niche. The third mode of environmental construction implicated by Lewontin in particular, namely that of undermining the conditions necessary for reproductive success, is not part of the concept of niche construction. Nor will all aspects of niche construction involve construction in a material sense. There are subtle distinctions to be made, and I will return to Godfrey-Smith (1996a) in the next section to elaborate on them.

For a start, however, it will be sensible to assume that organisms, in manifold ways, contribute to the environmental conditions under which they thrive – or fail to thrive. The activities involved need not always be adaptive: by altering the environment, animals change the conditions to which they will have to react further down the timeline, and sometimes these changes might actually be for the worse. One of the puzzles Lewontin sets out to solve is why, if adaptation by natural selection plays such a prominent role in evolution, organisms are not optimally adapted to their environments. If the conditions in some population's environment remain largely stable over time, one would expect that some state of optimal adaptive fit will be reached in many instances – if only conditions remain stable for long enough. Besides environmental pressure from competitors and other externally induced changes in environmental conditions that work against long-term stability, part of both the agency of change and the blame for sub-optimality may lie with the organisms themselves.

The aspects of specificity of an environment to an organism and of the bi-directional nature of the organism-environment dynamics are probably best integrated in an approach that, among other strands of heterodox biological theorising, incorporates environmental and niche constructionism, and that has become known as developmental systems theory (abbreviated DST and introduced by Susan Oyama 2000). Being a deliberate change of perspective on the animate realm rather than a conventional predictive theory, as Oyama et al. (2001, 1f) admit in the introduction to their developmental systems anthology, it reverses the direction of change taken by Richard Dawkins (1999) in his "gene's eye view": Dawkins suggests a perspective of inquiry under which the organism ultimately becomes "transparent" (1999, 4f, 250), so that what becomes visible instead are replicating gene sequences of which the organism and his environment in conjunction are the wider environment in which it acts, and which it manipulates. Thereby, the notion of the gene as the basic unit of natural selection shall be defended.

In contrast to this paradigm of an adaptationist view, a developmental system is introduced as a theoretical concept to comprise the conjunction of organismic and environmental factors that accounts for the presence of certain phenotypic traits within an organism or population, where environmental and non-genetic organismic factors are considered as intrinsic to the development of the organism as genetic ones. On the one hand, an identical set of genes might be found in clearly distinct phenotypes. For example first-generation worker ants of a newly founded colony, one of whose tasks is to construct many of the standard features of ant colonies, will look and behave differently from genetically identical later generation specimen raised in the fully established colony, and hence in an environment that was shaped by those first generations (Gordon 2001). Similarly, the sex of turtles and crocodiles is not genetically determined but depends on environmental temperature during embryonic development (Bateson 2001). On the other hand, modification of environmental factors may affect the development of traits that match with environmentally unmodified genetic variants, hence providing for distinct developmental routes to a similar phenotype, as in the phenomenon of "phenocopying" (Goldschmidt 1949).

Either way, the focus of inquiry has to be on the combined system, with environmental factors fully integrated but not necessarily interchangeable with genetic ones, and with the key unit of analysis often not being the individual organism but supra-individual entities. The very notion of one central unit of control in development might be misguided to begin with. Even mechanisms of inheritance might be distributed over a variety of factors, including persistent structures in the environment. There is no such thing as genetic information that could be taken by itself and still be informative about what phenotype an organism will develop. Such would be the case only if "strong instructionism" were true, that is if genetic information were supposed to fully specify phenotypic traits (Wheeler and Clark 1999; 2008). Rather than being relegated to the status of context, however important, to the content of the genetic code, environmental and other non-genetic factors are placed on equal explanatory footing with respect to informing biological development. Some developmental systems theorists even reject the notion of genetic information altogether, and speak of genetic and non-genetic inheritance of developmental resources instead (Griffiths and Gray 2001). One may prefer, but will not need, to adopt this latter view in order to appreciate how complex, dynamic and bi-directional the relation between organismic and environmental factors in development and evolution is. One may actually wish to retain the notion of genetic information in order to appreciate the interplay between information as specification and the specificity of organism-environment relations that often makes the former appear as underspecifying.

Construction and constitution

The ambiguity in Lewontin's argument among activities of construction in a literal sense, general modification of conditions by organisms, and the tracking of local conditions may have a very interesting upshot for my general argument. So it may be worthwhile to spell out the nature of construction involved here in some more detail. A critical inquiry of this kind has been undertaken by Godfrey-Smith (1996a, Chapter 5) in his account of the function of cognition as a means of coping with environmental complexity. He managed to distinguish between no fewer than five notions of construction in Lewontin's work. Construction may simply refer to modification of conditions in the external environment, where these changes are observable and measurable. It may, however, also involve the creation of properties of the environment that would not exist, *sensu strictu*, in absence of an organism's behaviours, where some of the properties are not observable and measurable in the same sense as changes to the external environment. Hence, organisms may be involved in activities in which properties of the environment are either *constructed* or *constituted* by their behaviours (see Godfrey-Smith 1996a, 144f).

In Godfrey-Smith's terms, "construction of the environment exists whenever an organism intervenes in formerly autonomous physical processes in the external world, changing their course and upshot" (Godfrey-Smith 1996a, 145). A formicary, a fox earth or, most paradigmatically perhaps, a beaver dam are features causally and materially constructed by organisms in a quite straightforward sense, and will be relevant to the respective animals' reproduction. Godfrey-Smith's take on construction, however, does not account for one sort of case of direct physical modification that Lewontin had in mind, namely situations where organisms overpopulate or pollute their environments, or otherwise causally undermine their livelihood and hence end up constructing their environments with reverse signs, as it were, in ecological and selective terms. Only physical changes that do have or may acquire an adaptive function for the organism and that may hence be subsumed under an evolutionary strategy for the organism are counted in by Godfrey-Smith, and only modifications that add structures to the external environment or alter its overall shape in relevant ways will count as construction proper.

A somewhat more abstract but equally important relation is established when organisms constitute elements of their environments: "Features of the environment which were not physically put there by the organism are nonetheless dependent upon the organism's faculties for their existence, individual identity or structure" (Godfrey-Smith 1996a, 145). An organism who brings himself into a relation to some environmental variable so that it suits his needs, by moving or otherwise behaving towards that variable, will be the paradigm of constitutive relations. On the physical side, the effects may well be negligible but, as in the case of a herd's territory, much in the life of the population depends on treating or perceiving certain features of the external environment in certain ways. A herd's territory, which, although having a spatio-temporal extension and sometimes being bounded by topographical features, too, will be sufficiently defined only by the herd's population dynamics, patterns of spatial movement, interaction with other populations and so forth, and the extension of that territory will alter in accordance with changes in these factors. A slice of space-time becomes an animal's or population's territory only by virtue of the animals behaving towards objects and organisms within that slice in a certain way while behaving differently, or not at all, towards objects outside that slice. The territorial animal acts upon the relation between the nature of an object or of another organism he encounters and their relative placement inside vs. outside the slice of space-time that is treated by the animal as his territory. The territorial animal hence deals with domain-specific regularities as outlined in NI-5 in the previous chapter, and doing so will be part of creating and maintaining his ecological niche.

The common denominator of constituted features of the environment, which include a variety of other ecological relations, such as being a mate or a competitor or a specific microhabitat, is that they can only be relationally defined. The physical nature of the objects and organisms inside and outside the slice of space-time that an animal treats as his territory does not make the difference in terms of relevance to the organism, apart from their relative placement inside vs. outside of that slice. Their being placed inside vs. outside the territory is not an intrinsic property of those objects and organisms but a relational property that exists only in view of the behaviours of the territorial animal. Where a beaver dam is a material structure with a function to the organism, a territory, mate or competitor exists only in relation to the organisms involved in treating other things and beings under that relation.[2] Similarly, a microhabitat, such as the crevices in the case of the arid-environment *Drosophila pseudoobscura*, although defined by local variance in physical conditions, does exist as that microhabitat only by virtue of being treated by the organism in a certain way, namely by virtue of his activity of tracking the variance and invariance in local conditions and moving where conditions are suitable.

In the light of these observations, it might seem that at least some elements of environments are indeed subjectively defined. As for natural information, this would be a premature conclusion though. That the existence and shape of territories have the presence and the behaviours of organisms as their necessary precondition does not make these entities restricted, ontologically, as it were, to the organisms' particular perspective. What they are restricted to is the domain of ecological and selective environments, and these can be described only in relation to the organisms inhabiting them. If a population moves to a different place, for example in response to depletion of resources, it will establish a new territory that will be demarcated by its members' behaviours, which will possibly include the use of some signals to the effect of marking the boundaries of the territory. Hence, environmental entities constituted in such fashion *are* independently observable, although by different means than direct physical modifications, namely by proxy of the behaviours of the organisms and their effects.[3] In this respect, observing and describing constitutive relations to an environment parallels observing and describing intrinsic information in perception.

Godfrey-Smith's distinction between organisms constructing and constituting features of the environment, although not identical with the distinction between physical and informational properties of the environment, stands in an instructive relation to the latter. Taken by themselves, physical conditions directly affect behaviour, and are directly affected by it. Completion of a beaver dam will quite directly improve conditions of success for the beaver's foraging behaviour, and hence will act positively on his reproductive chances. Cattle and beaver may be able to collect information on the physical changes they purposefully or inadvertently brought about. Some of the world affairs they encounter will then serve as signals of these changes. The animals might even be aware, to some extent, of the effects of these changes on themselves. Although the ability to use information on self-induced physical changes in the environment – on the possibilities and the consequences – is common currency among higher animals, there is a necessity for doing so only in a subset of cases. The animals may fail to grasp the changes they induce and still get along well (although not in the case of overgrazing cattle, but these are not constructive activities in Godfrey-Smith's sense anyway). In contrast, the constitution of features of the environment is always informationally based, because it by definition involves treating these features *as* something in relation to the organism himself, for example as being part vs. not being part of one's territory. The degree to which the uptake of information is involved in either case depends both on the number of variables an organism has to rely on and the degrees of variance in these relevant variables.

Not least for their orthogonal relation to the kind and amount of natural information involved, constitution and construction are strategies that are not correlated with degrees of organic complexity, let alone a presumed hierarchy in the tree of life between primitive and advanced organisms. Although building artefacts, on the face of it, may look like more of an accomplishment than moving to where the conditions are right or treating an area as a territory, many higher animals, too, will be found to follow the strategy of constitution rather than construction in order to cope with the vagaries of their environments. As examples, Godfrey-Smith's contrasts cockroaches, hawks and dolphins as followers of the constitutive strategy with termites, beavers and human beings as constructors (1996a, 146). Animals of different degrees of complexity in organic and behavioural traits may follow either strategy.

Construction and constitution are distinct strategies, and they will stand in various relations to the use of natural information, from information-extensive to information-intensive. This distinction should, however, not lead us to the conclusion that these strategies were mutually exclusive. We are much more likely to find differences of degree between them, with few organisms populating areas near the extreme ends of the spectrum. Conceptually useful as Godfrey-Smith's distinction between constitutive and constructive relations is, it should not misguide us into believing that we will find an equally neat distinction when empirically describing the behaviours of organisms. We are much more likely to find them combined in different arrangements across the entire spectrum of organic diversity and complexity. By implication, there will be instances of activities that include constitutive and constructive elements in conjunction.

An animal who marks his territory might not only be picking up natural information related to physical conditions that bound his territory. He may also be placing physical markers in the external environment, so as to make them perceptible to conspecifics who will then adjust their behaviour to the information they encounter. The features constructed by one animal may hence play a constitutive role for the behaviour of another. The effect will possibly be symmetrical, in terms of adaptive functions, while the process is not, as different activities are involved on the sender and receiver sides respectively. We might also encounter the condition that neither un-constructed information nor physical markers alone are sufficient to provide the required information. Only the two in conjunction will be sufficient. Although the effect of what is constructed might be material in an immediate sense, for example in physically keeping the other animal off one's territory, it is more likely to be mainly or wholly informational.

Similar conditions will hold for any population where, besides the uptake of natural information by individuals, signals are purposefully created and transmitted between individuals in accordance with some adaptive function, such as for warning cries or mating displays. For many animals, the making of their own ecological and selective environments involves the production of structures and events that occupy an intermediate position that integrates aspects of physically constructing environmental features and constituting specific relations between themselves and existing features of the external environment. Any active signalling within populations and any creation of artefacts that store and provide information for other members of some population, but also any signalling with deceptive intent to some potential prey or predator that deteriorates the receiver's informational situation, will characteristically occupy this intermediate position. If and when that signalling involves learning and some kind of accumulation and transmission between individuals in a population over time, not only the construction/constitution distinction is blurred in an informative way, but also the distinction between ecology and history of some population, in making information that was present in the past available and relevant to the present, independent of what physical conditions may matter in a given situation.

Notes

1 Most of the controversy between Gould, Lewontin, Vrba and their main opponents (Dawkins 1999, Chapter 3; Dennett 1995, Chapter 10) does not concern whether phenomena such as spandrels and exaptations actually exist but what implications their admission to evolutionary processes has: do they invalidate or only require us to amend the hypothesis that adaptation by natural selection is the key to evolutionary explanations?

2 The distinction at issue here is not entirely unlike the ontological distinction in the social studies of science between the construction of artefacts and the social construction of institutions, such as marriage, money or courts, as "realities created by references to them" (Bloor 1996, 842). The difference, of course, is that the latter realities are created in the course of shared linguistic practice, and thus on a higher level of informational complexity.

3 A similar point is made by Sterelny and Griffiths (1999, 269) who note in their discussion of Brandon (1995) that it is an *empirical* truth that organisms may modify their external environments, whereas it is a *conceptual* truth that they modify their ecological and selective environments. These latter kinds of environment are always determined in relation to the organism – by definition, as it were.

6 What is an informational environment?

Before providing more substance to my preliminary suggestions on purposefully produced signals and informational artefacts in Part II, it will be important to systematically match the account of organism-environment relations outlined in the preceding chapter against the account of natural information developed in Chapters 2 through 4. How *is* an environment informationally specific to an organism? How does the specificity of informational relations match with situations of underspecification?

In order to at least point towards an answer to these questions, I will now work towards a proposal that is based on a definition of what I will call "informational environments". This term was first used, to my knowledge, by Kim Sterelny (2003), with a partly similar meaning but, as we will see, within the context of a partly different argument. Keeping track of some of the similarities and differences to his account will be instructive to the present subject matter. Moreover, it will be useful to accord my account of informational environments with the conceptual distinctions between external vs. ecological (vs. selective) environments and constitution vs. construction introduced in the preceding chapter. Hence, in conclusion of Part I of this book, I will prepare and then present a definition of informational environments, and take a look at what happens when informational environments happen to change or are actively changed.

Environmental information and the use of cognition

In the previous chapter, I referred to Robert Brandon (1995) as identifying ecological environments, being the environments of main interest here, as the set of conditions generally relevant to an organism or population in terms of growth and reproduction. I also mobilised Richard Lewontin (1982), F. John Odling-Smee et al. (1996) and developmental systems theory to argue that ecological environments and, more specifically, ecological niches, are partly of an organism's and population's own making. Last, I referred to Peter Godfrey-Smith (1996a) and his more specific account of how organisms and populations enter into relations of constructing and constituting aspects of their environment and thus modify the set of conditions that are relevant to them, ideally to their benefit.

At first blush, none of these accounts seem to include a role for information to play in determining the set of relevant conditions or to be part of these conditions. With respect to the distinction among external, ecological and selective environments, this observation is correct, yet the exclusion of information on this level does not amount to an omission. Instead, it warrants a conceptual generality that is fully adequate: many organisms neither are able nor need to collect much information on many or most of the relevant conditions in their environment in order to meet their adaptive needs. To an external observer though, the traits and demographic performance of those organisms may provide information on conditions in their environments, in terms of covariance with the values of some relevant variable or variables in that environment. Conversely, the definition of an ecological environment, in terms of the set of conditions relevant to an organism, is impartial to whether and how information is collected about these conditions by its inhabitants. What matters first is *that* the respective conditions are relevant.

With respect to the construction/constitution case, Godfrey-Smith actually does not omit information from his account of the relation between environmental complexity and the function of cognition of which that distinction is part. Instead, he reserves his use of a vocabulary of information for what can be modelled in terms of Lewis- Skyrms style sender-receiver games (Godfrey-Smith 2013a; 2013b; see the discussion in Chapter 2). Natural information will not fall within the extension of this more restricted usage, where "unsent signs" figure as "cues" instead. Terminological preferences aside, Godfrey-Smith is duly concerned with the role of what I have been referring to as natural information (aka cues) in environments as a constituent of cognitive processes. His view echoes Fred Dretske's theory of information in some respects:

> Cognition is useful in an environment which is characterised by:
>
> (i) variability with respect to distal conditions that make a difference to the organism's well-being, and by
> (ii) stability with respect to relations between these distal conditions and proximal and observable conditions.
>
> [. . .] Cognition is most favored when there are (i) environmental conditions salient to the organism, which are not directly observable, and which are not stable or predictable in advance, and when there are also (ii) highly reliable correlations between these distal states and states which the organism can observe or detect more directly.
>
> (Godfrey-Smith 1996a, 118)

It is the "highly reliable correlations" between distal and proximal affairs that count as information in this context, where the distal affairs are the conditions at the source and the proximal affairs are the signals. Variability at the source is as important to the role of information in cognition as stability and unequivocality of the correlations (for the terminology used in this reconstruction, see NI-4 and its sub-clauses in Chapter 4). Variability at the source is what makes information relevant to the receiver. The set of relevant variable conditions at the source and the likeliness of signals to occur when certain conditions at the source obtain determine the reference class for the informational relations. On this account, high variability but not randomness at the source and little variability in their correlations to proximate, signalling states are what provides a function for cognition.

On the view presented by Godfrey-Smith, an environment may be too dull to warrant any cognitive effort on the one hand, in terms of the relevant distal conditions being too stable and too predictable to provide sufficient grounding for any adaptive function for mechanisms of memory, learning, inference and so forth. Dretske's magnetotactic bacteria may serve as an example of organisms living in such a dull environment. There is only one variable that is informationally relevant to the bacterium, and the correlation between the direction of the Earth's magnetic field and the gradient of oxygen concentration in marine waters is extremely stable. Unless, however, the bacteria were simply being pulled downward by magnetic forces instead of the organisms actively following the lines of the magnetic field as a proxy of the gradient of oxygen concentration, there will be information involved in tracking that one relevant variable. Still, the evolution of cognitive abilities will require more than that.

On the other hand, an environment may also be too chaotic to warrant any cognitive effort if distal conditions change randomly or if there are no reliable correlations to proximal affairs that the organism could detect. There would be no information for him to use in cognitive processes. Here, sheer resilience in organic functions or the ability to reproduce in large quantities whenever conditions are favourable will be workable strategies. This is as close as it gets for life to being information free. According to Sterelny (2003, 149), this kind of case forms a class of its own, besides "tracking" favourable conditions (the constitutive strategy in Godfrey-Smith's terminology) and "ecological engineering" (the constructive strategy). This is the strategy of "adapting" to conditions in the environment, which seems to echo that classical interpretation of adaptation in evolutionary theory on which environmental pressure works on lineages of otherwise inert organic matter. Here, however, it is only one class of strategies among others to create an organism-environment fit, not the expression of a natural law of natural selection.

Conditions will be neither too dull nor too chaotic for any animal that uses natural information in cognition. It would look reasonable to assume, within Godfrey-Smith's conceptual framework, that purposefully created signals might be used in the same fashion. If the function of the mind, on his "Environmental Complexity Thesis" (1996a, 3) explicated in the previous quotation, is to "enable the agent to deal with environmental complexity", and if environmental complexity is understood as the heterogeneity and variability of conditions therein, purposefully produced signalling events or artefacts may serve as an environmentally rooted means of dealing with a given situation of complexity in a manner analogous to natural information, as long as they fulfil the second of the aforementioned conditions, namely that of stability of relations between proximal and distal conditions.

There might or might not be a competitor claiming that pasture over there as his territory. His markings will tell me, even if I cannot see him. There might or might not be a predator around the corner. Another individual's warning call will tell me, even if I cannot see that predator myself.

There are two apparent reasons why Godfrey-Smith does not consider the possibility of an analogy of functions between natural information and purposefully produced signals: first, his concept of cognition is explicitly restricted to a pragmatic interpretation in the Deweyan tradition, where it figures as a problem-solving capacity possessed by individuals in the first place. Perceptual cues that directly guide action in a Gibsonian sense are, on Godfrey-Smith's view, no natural component of this picture of cognition. This, however, might not be the only or the best way of relating Gibson's work to the tradition of American pragmatism of which Dewey was part. Both Anthony Chemero (2009) and Harry Heft (2001) draw a fairly direct genealogical and systematic line between the "American naturalists", particularly and respectively James Dewey and William James, and James Jerome Gibson's ecological psychology. Gibson's conspicuous eschewal of mental representations as explanantia in an empirical psychology is cited as a key testimony to that heritage. At the end of Chapter 7, I will suggest that Dewey's notion of the organism-environment relation offers another analogy to the Gibsonian view. According to Godfrey-Smith's view of cognition, Gibsonian perception-action will be a worthwhile strategy only in an environment that is distally relatively more dull than, for example, a human environment, but still proximally reliable. One part of my argument, however, was to maintain in a Gibsonian vein that much of human activity is just as directly guided by perception as that of cognitively more humble animals. This does not amount to dismissing the importance of the additional accomplishment of internal sender-receiver games that constitute cognition, and Godfrey-Smith's merit lies in describing the conditions under which cognition, in the traditional narrow sense (no pejorative connotations intended for either predicate), is valuable. However, maintaining that direct perception is essential to human action, too, implies that only *some*, however important, aspects of human environments provide the right sort of distal-variability/proximal-reliability ratio to ground a set of functions for higher-order, narrow cognition.

Second, if cognition is defined as an internally based problem-solving capacity on the background of the kind and level of environmental complexity an organism has to deal with, the possibility will not be acknowledged that the accomplishment of some of the tasks of coping with environmental complexity, and hence, in Godfrey-Smith's terms, cognitive tasks, may be delegated to the environment by means of informational artefacts that still remain constituents of cognition proper. I will discuss the possibility of constructed entities in the environment that play this latter kind of role in cognitive processes in more detail in the next two chapters (Chapters 7 and 8). It will suffice to say for now that, even if one sticks to the view that cognition should be considered as a set of internal, neurally based processes in the first place, there are good reasons to believe that creating signals and placing them in the environment does amount to a kind of externalisation of the accomplishment of cognitive tasks into the environment, especially in view of the fact that environments are normally shared between individuals and require coordinated activities.

A third and possibly stronger reason for assuming that signalling within an organism and the purposeful placement of signals and informational artefacts in the environment are disanalogous affairs is provided by Sterelny (2003, Chapter 2): the production of signals by an organism, especially when directed towards organisms outside one's own population, may not always and not even typically serve cooperative purposes. Although the uptake of natural information on physical conditions in the environment is a straightforward affair, Sterelny argues, everyone who, like the proponents of behaviour-based AI, focuses on this kind of information misses the importance of signals that are made with the purpose of deception. A potential prey may use mimicry in order to trick a predator into mistaking him for a poisonous or otherwise unpleasant animal, or a predator may use camouflage in order to making him being mistaken for a piece of a plant or otherwise benign object in the environment. In such cases, the function of the signals or structures in question is to actually *reduce* reliability and unequivocality of the information available to the receiver, and hence deprive the signals of their informational, that is specifying character. Only if and when means of finer discrimination or additional cues are or become available to the receiver, can the signals in question be disambiguated.

Hence, Sterelny concludes (in explicit reference to Godfrey-Smith 1996a), one key function of cognition is to match less-than-reliable cues encountered in a multifarious, "informationally translucent" environment against each other. Precognitive "detection systems" that are confined to one or a few non-integrated information channels will be a viable option for organisms who are relatively more robust and specialised in their constitution and, as the flip side of the same coin, live in less variable, "informationally transparent" environments than those generalists for whom a complex array of conditions will matter and who have a rich repertoire of responses to match variation in their environments. For the latter, "robust tracking" will be the best solution. An "informationally opaque" or, in the terms referred to earlier, chaotic environment will, in terms of adaptive economics, not warrant that effort but favour fixed responses and resilience. The environment whose informational characteristics are most suitable to the evolution of complex cognitive functions, as Sterelny argues, is neither transparent (or dull) enough to always and in all respects provide reliable cues nor opaque (or chaotic) enough to automatically frustrate any cognitive effort.

What informational environments are

On the background of the preceding discussion, let me now introduce an account of informational environments that caters for the various shades of informational characteristics of environments as outlined by Sterelny (2003), from opaque to transparent, while reserving the character of being genuinely informational to those signals which actually serve an adaptive function for its receivers. In departure from the distinction between natural cues and sender-receiver games proposed by Godfrey-Smith, my account shall equally cover natural and purposefully produced signals, to the extent that the latter have a bearing on their senders' and receivers' ecological environments. That bearing typically – but not always – consists in serving purposes of coordination and cooperation within a population of users, that is senders and receivers of such signals. Situations of competition and deception will constitute a relevant special case. In terms of a definition, my account will assume the following shape:

(IE-1) The informational environment E of a receiver R is the subset of the proximal events and objects present in his external environment that *I*-regularly covary with *F*-conditions at a distal source s and can be detected by R, as outlined in NI-1,. . ., 5 (see Chapter 4), where

(a) the *F*-conditions at s signalled by members of a type **r** within E are in a broad sense ecologically relevant to a population of individuals $\{R_1,\ldots, R_n\}$ of which R is a member, so that the detection of **r** serves an adaptive function to them;

(b) the relations between tokens of **r** and *F*-conditions at s are, globally or within domains D, either (i) stable and unequivocal as such, or (ii) can be disambiguated to a sufficient degree by members of $\{R_1,\ldots, R_n\}$, provided that ecological conditions in their environment remain within a normal range;

(c) the degree of variance in reliability of information according to IE-1b accounts for the relative richness vs. impoverishment of an informational environment with respect to R's constitution and abilities, provided that detectability conditions remain above the threshold described in NI-4d.

(IE-2) E may contain types of purposefully created signals $\mathbf{r}_a$ whose emission or detection serve adaptive functions for a population of cooperating or coordinated $\{R_1,\ldots, R_n\}$, where

(a) type $\mathbf{r}_a$ signals are produced by individuals within $\{R_1,\ldots, R_n\}$ and thus are constructed aspects of the environment that can be used towards constitutive purposes with respect to *F*-conditions at s;

(b) type $\mathbf{r}_a$ signals can be used by members of $\{R_1,\ldots, R_n\}$ *either* (i) in analogous fashion to natural type **r** signals, as in IE-2a, *or* (ii) will introduce new mapping regularities or new kinds of sources $\mathbf{s}_a$ or *F*-conditions at s, with detectability conditions as outlined in NI-4 having to apply;

(c) there is a possibility that members of $\{R_1,\ldots, R_n\}$ produce signals of a type $\mathbf{r}_d$ that have as their adaptive function not the provision of information to members of that same population but interference with the informational situation for other, non-cooperating $\{R_1',\ldots, R_n'\}$.

(IE-3) Conditions within E might be partly controlled and modified by members of $\{R_1,\ldots, R_n\}$, so as to maintain or create stability, unequivocality and detectability conditions for type **r** signals.

An organism's informational environment is thus made up of those proximal signals, natural or artificial, which, under normal conditions, relate to distal conditions that are ecologically relevant to him. It is the stability and unequivocality of the informational relations between what he receives as signals and the variable conditions at the distal end that enable him to deal with the relative uncertainty on that latter side. That stability and unequivocality, however, cannot be taken for granted, as both ecological and informational conditions in his environment may be subject to change, and as they may be so independently from each other. In analogy to ecological environments, informational environments are specific to an organism's constitution and abilities. Even if and when external environmental conditions are identical, informational environments will be at variance between differently disposed organisms. The information present in the environment will be taken up differently, by different means, and will be relevant in different ways and acted upon differently. The activities will include the possibility for organisms to modify informational environments in accordance with their specific needs and purposes.

Under what I called "normal" conditions in IE-1b – which are to be understood in Ruth Millikan's sense as conditions that are necessary and partly sufficient for an explanation of the nature and the functions of whatever is being reproduced in a given context (Millikan 2004, 33f) – the informational environment is made up of those informational relations which, qua being sufficiently reliable proxies of relevant conditions at the source, actually *are* relevant to an organism. They have assumed their status as relevant over the course of a history of interactions between organism and a subset of the ecologically relevant conditions in the environment that involved and turned out do depend on the presence of signals of a given type.

The ecological relevance mentioned in IE-1a is characterised as "broad" because no direct effect of the conditions at the source or the presence of some signal on R's reproductive chances is required. The effects may well be indirect, in terms of informational conditions affecting the success rates of responses to conditions at the source that have an effect on the reproduction of the behaviours themselves rather than the reproductive chances of its bearers. The paradigms of this kind of situation include all instances of practices that are transmitted between individuals and hence reproduced by means of observational learning, and thus all sorts of phenomena investigated under the "cultural evolution" or "dual inheritance" monikers (Boyd and Richerson 1985; Cavalli-Sforza and Feldman 1981; Tomasello 1999). Behaviours may be directly replicated by imitation or reproduced in view of their purposes in what is called emulation learning. Either way, their reproduction is subject to mechanisms of selection of its own, whose primary target are the distal conditions to which those behaviours relate. Hence, many things that are not a matter of life or death and, by consequence, many things that are not selectively relevant in a strict biological sense, may be relevant to some organism or population in less dramatic and more indirect but nonetheless important ways – which, if and when effects persist and accumulate, might have implications for biological evolution.

The difference between an organism's ecological and his informational environments is this: an ecological environment is constituted of those variables and their respective ranges at the distal end that are relevant to an organism. It is possible but not necessary that the organism can also detect the values of those variables. He might be built for resilience or fast reproduction instead. An informational environment, in turn, is constituted of those proximal affairs, naturally present in the environment or produced and reproduced in behaviour, which stand in sufficiently reliable relations to a subset of the ecologically relevant conditions at the distal end, so as to enable an organism to keep track of those distal conditions. The distal affairs may be either spatially, temporally or perceptually remote from the organism, and in some cases their relevance may be similarly decoupled from the here-and-now. Both may lie in the past (this is what did *F* to me), in the future (this *F*-condition will return) or be merely possible (I might be in trouble if this *s* turns *F*).

One decisive criterion for the inclusion of information present in the environment as part of an organism's informational environment is that the receiver of that information is able to respond in some way to conditions at the source (see IE-1a) – either by tracking favourable and avoiding unfavourable conditions (Godfrey-Smith's constitutive strategy) or by altering these conditions (the constructive strategy). Still, there will be distal conditions that are ecologically relevant to an organism, but on whom he cannot collect information, either for want of stable correlations between proximal and distal affairs, or for want of means of collecting that information, or for want of means of responding to it. Such might be the case for what I called background conditions on p. 81 in Chapter 5: conditions on whose relative persistence an organism relies, but which may be beyond his means of access, informational or other.

Conversely, not all information in a receiver's environment is part of his informational environment, even if he is able to receive it. If the presence or absence of some information does not make any difference to his behaviours in general or to the achievement of his goals in particular, not even at the most distal reaches of what he can do in this world, that information does not have an adaptive function to him. It could just as well be indifferent, signal-free noise, unless it happens to interfere with his activities in other ways. Its status may, however, be subject to change. If an organism stumbles upon a *prima facie* useless bit of information, it might turn out to make a difference at some later point in time.

Besides natural information that is relevant and available to its receivers, an organism's informational environment is likely to also contain information that has been purposefully added by the organism himself or other organisms, as indicated in IE-2 and its sub-clauses and briefly discussed with reference to the reproduction of behaviours. In terms of the correlations involved, the conditions of transmission and reception, there is no principled difference between purposefully produced and natural information. However, there are differences, first, with respect to the correlation or "mapping" between signal and source not being warranted by lawful or statistical regularities in nature but by regularities in usage or, under some circumstances, definition. Second and relatedly, instances of a signal not mapping onto its source will have to be evaluated with respect to the sender's underlying purposes, as these may include production errors (i.e. malformed signals) and errors of mapping (i.e. misplaced signals) as well as instances of intentional non-mapping (i.e. deceptive purposes).

As to the condition of the regularities of mapping, the relation between signal and source is determinate for natural information, as the signal's presence and shape could not have been otherwise under a given set of conditions, whereas it is indeterminate for purposefully created information. The signal may or may not be chosen to mimic natural signals. The signal may be chosen to depict its source, to bear a more abstract symbolic relation to its source, or simply to identify its spatio-temporal location. What unifies all kinds of signal-source relations is that they are, first and foremost, relations of mathematical mapping, as explicated on p. 67–68 above. As such, they can be abstract and indirect in manifold ways. A *q*-bird's track in the woods physically matches and, in this sense, resembles the shape of the bird's feet, whereas the orientation of magnetic particles in magnetotactic bacteria, although faithfully mapping onto the gradient of oxygen concentration in marine waters, does not bear a relation of resemblance to the source that would be detectable either for the bacteria themselves or for the human observer. The only relevant variable to be mapped is that of direction. An arrow and an array of symbols used on a signpost are supposed to map the identity and relative location of a restaurant, a cashpoint or a restroom. In conjunction, they will be the relevant variables for variantly disposed users (Do I need food, cash or a restroom *now*?). A description of the location of each of these entities from a local person, if and when accurate, will include the same mapping relations.

As to the condition of the status of instances of non-mapping, if a signpost, description or other purposefully created signal are malformed, erroneous or deceptive, the mapping between signal and source will not obtain, without necessarily disqualifying the signal as an informational artefact. This is a condition that does not and cannot apply to natural information, no matter what its degree of reliability might be. As Dretske (1986), Millikan (2004) and others have pointed out, only things that have been designed to provide information can actually, and for various reasons, *fail* to do so. My weather report but not black clouds on the horizon may be mistaken about the imminence of rain. Still, black clouds may be a reliable-enough indicator of rain for guiding my response. Conversely, I might devise a signal with the purpose of not providing information, achieving my mischievous purpose if I deceive the receiver precisely by producing something that has the appearance of a properly formed and properly mapping signal. Only if and only when the possibility of deception arises, and hence only where signals are purposefully produced, cooperation and coordination within a population of signal users becomes an issue deserving closer scrutiny on the informational level. Hence, they are first mentioned in IE-2. Populations that exclusively use natural information may or may not act in cooperative or coordinated fashion, but the only way in which they can define or defy cooperation and coordination will be in revealing or concealing a natural signal to other individuals.

It is in light of these considerations on the normative nature of natural and artificial signals that some of the clauses in the earlier definitions should be read – which may have the additional effect of making the notion of the receiver-specificity of informational environments appear at least a bit less trivial than one might first believe. To begin with, the condition in IE-1a that, in order to be a constituent of an informational environment, information has to serve an *adaptive* function for its receivers and, by analogy, serve a function not only to an individual but also, potentially and in principle, *for all members of his population,* is designed to keep out arbitrarily formed signalling relations. Without this double specification, signals placed in an individual's environment would have to be counted as part of his informational environment that are designed and used to serve contingent, idiosyncratic and ad hoc purposes either for that same individual while not possibly doing so for members of his population, or that serve adaptive functions only for members of *other* populations.

If such signals were admitted, the first type of case would represent one aspect of Lewontian environmental construction that has been demonstrated to obscure the very distinction between just any effect that an organism's behaviour has on his environment and those effects which have a history of being useful to members of a population and hence acquired a proper function in Millikan's terms: there might be signals that are useful for one or a few individuals but have the intended or unintended effect of undermining conditions for the population of which these individuals are members. Speaking in game-theoretical terms, an informational free-rider may maximise his individual benefits, but his selfish ways cannot provide a basis for a dominant strategy of acting within a population, as the reproductive chances of its members in general would be jeopardised at some point – especially if coordinated activities are relevant to that population. Even where competitive or otherwise uncooperative relations between individuals within one population are concerned, universal deception and unreliability of information could not be the norm. There would be no use in even trying to pick up information then. For this reason, I refer to cooperation *and* coordination in IE-2: displaying signs of putative dominance in a group's pecking order or in fending off another group is not precisely cooperative but still serves coordinated actions within a population, nor is it deceptive. Coordination may occasionally be served by deception too, unless it becomes self-defeating on the group level.

In the second type of case, however, there is a perfectly adaptive function for signals that are emitted by some individual or individuals within a population in order to deceive a potential prey or predator, host or parasite, as indicated in IE-2c. The individuals to be thus deceived typically are *not* members of the signalling individuals' own population, and the signals in question are produced precisely in order *not* to provide information. Here, deception will be the norm when viewed from the sender population's side, whereas there certainly is no adaptive function of those deceptive signals when viewed from the receiver population's side.

Consequently, the informational environment of a receiver *R* does, strictly speaking, not comprise signals that are intended to deceive *R* in the same way as it does properly mapping signals, nor does it equally comprise signals that are too ambiguous to be useful to *R*. These signals are detectable for *R* but make conditions in the informational environment more ambiguous, "opaque" or "impoverished", *for want of bearing informational relations to a relevant distal condition.* On the positive side of the balance, *R*'s informational environment will contain any signal that helps him to make the distinction between natural and deceptive or otherwise ambiguous signals, and hence bears informational relations to relevant distal conditions. Making the distinction between signals that appear identical but relate to very different conditions at the source will matter to him. Although the detection of false negatives, that is signals that do not appear to be related to the same source conditions but in fact are, is relevant to be sure, detecting false positives will normally be of greater importance. You might live with detecting a predator when there is none around, and even with doing so quite often, but you simply do not want to mistake that well-camouflaged predator for an innocuous piece of wood.

Hence, *R*'s informational environment will at least become more complex and challenging if and when a deceiver or other unfavourable conditions enter. It will impose additional discriminatory tasks upon him if and when he is able to use additional cues for distinguishing between genuine information and the signals produced by deceivers, impostors and other sources of equivocation. If I may borrow Sterelny's terminology for my purposes, *R*'s informational environment will have become less transparent and to an increasing degree merely translucent. If there is no such option available to *R*, he becomes less likely to get things right about what happens at the distal end. Hence, the deceptive signal will actually *subtract* information from *R*'s informational environment and impoverish conditions therein (see IE-1c). *R*'s informational environment will have become less transparent and more opaque. However, only if the balance between properly informational relations and merely seeming signals is overall positive, in terms of maintaining *R*'s purposes to a minimally sufficient degree, will an informational environment remain intact, although possibly become impoverished. Should the appearance of a deceptive or *de facto* irremediably ambiguous signal interfere with trackability beyond that point, we might consider *R*'s informational environment having turned dysfunctional. He might not persevere after all.

At any rate, admitting for the possibility of probabilities of signal-source mappings to vary over space and time (see NI-5a in Chapter 4) does not amount to admitting for a variance in the underlying regularities but to acknowledging the possibility of variance in epistemic contact (Withagen and Chemero 2009, see p. 60–61) brought about by variance in ecological conditions, with the caveats I added to that discussion. The informational character of a signal is grounded in the type of sources and source conditions that bring it about – hence disallowing alternate sources or alternate conditions at a source as part of the same type of informational relation (see NI-1). If an organism with deceptive intentions enters *R*'s environment, the reliability of his detection of the relevant conditions in question and hence his epistemic contact with the environment may be negatively affected but not the regularities that connect the 'original' natural signal to its natural source. After all, the natural and the 'treacherous' artificial signal in such cases are related to very different sources by different regularities, and are indistinguishable from each other only to the extent that *R* has no other cues available that would reveal the difference to him.

Keeping cases of deceptive signalling apart from the genuine constituents of informational environments has a normative aspect: it honours the environmental grounding of the principled possibility for *R* to distinguish between things that are different in kind even if and when signals emanating from them are indistinguishable at first sight. If signals that are indistinguishable for *R* when using his normal means of information uptake indeed have very different aetiologies, there will be a point at which the variance in aetiologies is detectable in principle. In all realistically conceived cases, sparing Descartes's malicious demon, brains in a vat and their kin, there will be at least a chance for that variance to also become detectable in practice – however remote that chance might be. It will be difficult to conceive of a situation in which the variant aetiologies will informationally manifest themselves as being identical on all levels and in all possible ways. (Could Twin Water, or the designer of Twin Water, not only pretend that it has all the phenomenal properties of water but also maintain a pretence of that substance being H_2O in the most sophisticated of laboratories, except that it is *not* H_2O?) Hence, although deception is common in many animals' environments, and although the world may be hostile to its inhabitants in many ways, such hostility should not be considered the norm. In a perfectly treacherous world, there would neither be a possibility nor a use for getting things right. Conversely, in a perfectly benign world, there would be no need for an ability to tell right from wrong, because, apart from internally caused malfunction, there would be no risk of getting things wrong to begin with. The possibility and the usefulness of getting things right with some reliability should provide the normative grounding for steering through a world that often tends to wrong-foot us.

How informational environments change

The possibility of an informational environment to become impoverished in the way previously described in IE-1c is a condition that should not be confused with either of two other issues concerning the shape of informational environments: on the one hand, an impoverished informational environment is not equivalent to an informationally more primitive informational environment, as it would pertain to an organism who relies on fewer or less interconnected cues. We would then be comparing the informational environments of differently constituted organisms, not different states of one organism's informational environment. On the other hand, an impoverished informational environment should not be equated with an environment in which the regularities that connect source and signal have altered. In such cases, as I argued in the preceding section, something more fundamental in the organism's environment has changed. We will then have to consider the possibility that his ecological or external environment have been transformed in a significant way, with secondary effects on the informational conditions relevant to him.

Notably, IE-1c also implies that an organism's informational environment may become enriched, for example by artefacts that are designed to enhance detection of some relevant condition, by providing additional signals or by making the uptake of information more reliable. In analogy to the case of impoverishment, we should not be tempted to conclude that *R*'s informational environment has become more advanced in an absolute sense, nor should we believe that the regularities connecting signal and source have improved in terms of being altered from a vaguely probabilistic towards a more law-like state of affairs. Either way, the regularities underlying the 'original', natural signalling relation remain the same, and they will amount to a specification relation, although the probabilities of correct mappings between signal and source will have altered due to intervening factors, which add to or subtract from the quality of what is received.

Having thus distinguished alterations of the informational relations proper from alterations of informational environments, what more can be said on the conditions under which an informational environment may count as having changed? To begin with, whatever change there is and on whatever level it occurs, it must be *persistent* with respect to the receiver's timescale of observation in order to qualify as a genuine change. Momentary aberrations in channel conditions or transient fluctuations in mapping frequencies that the receiver can correct for or integrate, or that he might be entitled to ignore as being too insignificant altogether, will not count. Moreover, the change in question must be *consistent* with the regularities in his environment. Inverted spectra or green things turning bleen all of a sudden, as they do in some philosophical thought experiments, will not count either. If conditions in an environment change, and even if some of the regularities in that environment are affected, such change will hardly occur in capricious fashion, nor is it likely to merely affect isolated variables.[1]

In terms of its origins, a change in a receiver's informational environment may owe to a change in his ecological environment. It does not necessarily depend on an ecological change though, and if it indeed does so depend, the ecological change is what should be the main matter of concern. A change in an informational environment may, and often will, entirely occur on the informational level, while having possible repercussions on the ecological level. Such an informational-level change might be of basically two kinds, as indicated in IE-2b, allowing for all sorts of shades between them: first, the availability and quality of some of the information usable to *R* may vary over time and space, in terms of new channels or new types of signals being added, removed or beginning to interfere with each other. Second, however, novel informational relations, with their own mapping regularities, may emerge that had not been present in the environment before. On the account of natural information that I have presented in this and the preceding chapters, an addition of genuinely new informational relations to the environment will occur only if some of the natural regularities that govern the mapping between signals and sources are altered or new ones are introduced. Supposedly, this is an exceedingly rare case. Conversely, new kinds of sources of information may appear to which established regularities apply.

There is an obvious, even trivial sense in which new information emerges in our environments all the time. This sense should be kept clearly distinct from the second way in which, I suggest, informational environments may change: with every new dawn, there will be new information on the position of the Earth in the Universe and on the overall conditions on our planet. With every newborn child, new information arrives, both in terms of his or her ever-so-slightly unique genome coding for ever-so-slightly unique traits and in terms of him or her signalling, in ever-so-slightly unique ways, current metabolic needs or a desire to be close or not so close to mother or father. Even where the domains of natural information concern individuals, these individuals normally do not appear as isolated entities but as members of some population, species or other natural kind for which there are established types of informational relations, based on natural regularities that are not contingent on the individual. In contrast, the appearance of life on Earth will be a promising candidate for the novelty I have in mind here, and probably the best candidate one could find in nature: genetic and developmental information, their regularities and mechanisms of mapping onto phenotypical traits and their modes of transmission were genuinely new phenomena when they first appeared, as were the evolving species of organisms themselves.

Apart from the phenomenon of life on Earth, the most credible paradigms of the "new kinds of information" case will be found in artificial structures that not only produce new signals but also introduce new ways of mapping them onto their sources, and that are difficult to subsume under pre-existing natural kinds. Language will appear as an equally obvious and ambiguous example in this respect: will the appearance of grammatical structures be entitled to count as the appearance of novel regularities of informational mapping, or will language only add new channels to the uptake and transmission of whatever information there is? The latter option might be too limited, as some information that human beings encounter and use in their environments would not exist without the use of language in the first place, and as grammatical rules are a necessary precondition of the existence of some of the structures that will count as sources of information. There is a number of linguistic structures that not only embody specific mapping rules which would not exist outside of language but whose use is also constitutive to entities that can then be referred to by linguistic means. One might think of the various applications of the negation transformation and its range of semantic effects, or of performative speech acts, such as promises or orders. Other examples of the "new kinds of sources" kind of case will include formal languages or works of art that are supposed to refer neither to natural objects nor to their subject in any naturally established way of informational mapping.

Prima facie, it will appear a more straightforward affair to provide real-world examples for the "new channels" than for the "new kinds of information" kind of case, but there are some difficulties. First, consider the informational properties of measuring instruments and other observational methods in science, which are supposed to make new modes of taking up natural information available that could not be found in a natural environment while affecting neither the relevant distal conditions nor the mapping regularities involved. However, the boundary to a certain class of technological artefacts will be porous, where both mapping regularities and kinds of sources might be subject to modification. Laboratory-based scientific observations will be a case in point here, where not only the conditions of observation have to be methodically and carefully crafted but also their object and the way in which it provides information about natural affairs. Simulations of world affairs will provide another example of the difficulties in distinguishing between creating new modes of uptake of natural information and the creation of artificial structures that introduce new kinds of information: is the simulation intended to refer to a concrete world affair or to some kind of general, idealised set of world affairs, or even to a fictional one, and how is reference accomplished in these cases? I will discuss this issue in Chapter 8.

Leaving most of these interesting and relevant complexities of the status of artificially created information in science aside, the paradigm of changes in the informational environments of human beings will be the introduction of technologies that, without necessarily interfering with the ecologically relevant conditions in people's environments, alter the way in which information that is relevant to their activities is available to them, so as to alter the way in which they collect and act upon that information. In some cases, the kind of information available to them will have changed, too. These are the kinds of change that I will discuss in Part II in some more detail. The argument will look into two complementary directions: first, what do we do when we add something to an informational environment? We might be extending the reach of perception and cognition, but there are several meanings to what that extension is, or might be. Second, what happens when informational environments are changed? Perception and action might be affected in several ways, and strategies might be required for accommodating for those changes. The effects of the extensions in question and the strategies of accommodating them will be worth exploring by means of taking a closer look at some examples of current and emerging technologies.

Note

1 I have to thank Volker Munz and Harald Wiltsche (personal conversation) for highlighting this problematic aspect of an important subclass of philosophical thought experiments. Although the possible worlds and counterfactual conditions thus conjured up are problematic with respect to the conditions of real-world perception, the thought-experimental method as such echoes a basic understanding of experimental practice: selecting, isolating and manipulating some variable of the target and measuring the results. The analogy will be much more substantial for scientific thought experiments though.

Part II

Environments of intelligence

7 The extension of the extended mind

In Part I, I have argued in some detail for an intrinsic connection between an organism's constitution and abilities and the make-up of his informational environment. I have also argued for the essentially natural character of the information involved, as being information for perception, as conceived of in ecological psychology. Informational environments, I continued, are determined with respect to conditions of ecological relevance. At the same instance, I acknowledged the possibilities for an organism to modify his ecological and informational environments. Hence, if there are activities in which parts of the environment are modified in constructive or constitutive fashion – in keeping with the technical meaning of these terms introduced by Peter Godfrey-Smith (1996a) and discussed in Chapter 5 – and if these modifications have an effect on the shape and availability of information, cognitive processes will not be confined to somatic boundaries. They might actually depend on features of the environment that are used or made to serve cognitive purposes. The accomplishment – and in a relevant subset of cases also the establishment – of some of the functions of human cognition may ultimately rely on the presence and operation of things that are neither in our heads nor in our bodies. This proposal has a prima facie parallel in the "Extended Mind Hypothesis", as introduced by Andy Clark and David Chalmers (1998), which shall be explored in the present chapter.

In their brief, bold and controversial manifesto of a thorough, "active" externalism, Clark and Chalmers highlight "the active role of the environment in driving cognitive processes" (1998, 7). Their version of externalism goes beyond established externalist views in the philosophy of mind and language respectively (see Putnam 1975b; Burge 1979 in philosophical semantics and McGinn 1977 in the philosophy of mind), according to which the content of mental processes or linguistic expressions depends on what is in the environment. In contrast, Clark and Chalmers argue that there are situations in which "the human organism is linked with an external entity in a two-way interaction, creating a *coupled system* that can be seen as a cognitive system in its own right" (1998, 8, emphasis in original). This externalist position, in its general outlook, shall be endorsed here. However, the notion of cognitive extension has become many things to many people in a continuously sprawling debate, where the most forceful critiques appear to frequently miss the most serious points that arguments for extended cognition are capable of making.[1]

On pain of adding another specimen to the zoo of views of what cognitive extensions really are, I shall carve out an interpretation that strengthens the functionalist credentials of the Extended Mind Hypothesis by rooting it in natural history. My guiding questions will be these: first, how shall the functional coupling between the organism and some entity in his environment be spelled out in detail? Second, what are the paradigmatic external entities to enter into that coupling? In seeking to answer these questions, I will make a suggestion towards sharpening the notion of "extension" that is (or should be) in play here: the paradigm of cognitive extensions are those features of the environment which play either of two kinds of constitutive role – which has to be distinguished from Godfrey-Smith's notion of constitutive strategies, which will be disambiguated as "GS-constitutive" where required – to the function of some cognitive trait. In functional-historical terms, these features have been primarily natural rather than artificial. This suggestion shall help to prevent the notion of cognitive extension from becoming extensionally bloated to the point of vacuousness.[2]

I will begin by placing some of the central claims of the Extended Mind Hypothesis – its notion of "active" vs. "passive" externalism and its notion of features of the environment serving cognitive functions – in the context of aetiological theories of function and some positions within evolutionary biology. The insights from this discussion will then be used to identify a number of possible interpretations of what extensions of the mind are, and what explanatory roles they might play in cognitive inquiries, with particular attention to the specific ways in which features of the environment may be constitutive to cognitive functions. The constitutiveness theme will be further explored in the third section of this chapter, with the aim of providing some common ground for "parity" and "complementarity" views of extendedness. In conclusion of this chapter, I will discuss two paradigms of cognitive extension in the light of the aetiological variety of functionalist argument developed before. In Chapter 8, I will explore artefact-based modes of cognitive coupling in more depth and match this account against the notion of informational environments introduced in Chapter 6.

The extension of functional histories

When locating themselves within the context of contemporary philosophical debates, Clark and Chalmers (1998, 9) characterise active externalism as "active" for its focus on interactions between organism and environment in the here and now. The activity of interest is in the environment and the organism at the same instance, and it is that concurrent activity which serves to make both cognition extended and externalism active. Conversely, Clark and Chalmers characterise traditional semantic externalism as "passive" for its focus on the causal history that supposedly endows linguistic items and, by extension of the argument (McGinn 1977), mental events with their meanings. It is this history of interactions that explains any possible difference between the contents of two prima facie identical mental or linguistic tokens, no matter what the current interactions may look like to participants and observers.

Hence, if one takes active externalism by its ahistorical first word, the chemical difference between water and twin water in Hilary Putnam's well-known thought experiment (1975b), contrary to what its author aimed to demonstrate, will *not* matter as long as my interaction with either substance produces identical effects, cognitively and practically. Conversely, on the "passive" account, the different histories of earthly and twin-earthly water are all that matters in making the difference between the semantic success conditions for "water" tokens being applied to either substance, even if I cannot see, feel or taste any difference between the two.

Although the active/passive terminology is now well-established (see, e.g. Carter et al. 2014), there remains a certain artificiality to referring to the varieties of externalism in this way. Past interactions have been relevant to shaping the content of some present linguistic or mental token, and there is no reason to assume the past to be in any normative sense remote from the present. What Clark and Chalmers are after is quite different from Putnam-style semantic externalism though: their focus is on the locus of cognitive processes, whereas Putnam, Burge, McGinn and many others are concerned with the external conditions that ground the content of mental or linguistic tokens. The difference at issue here is also known as "vehicle" vs. "content" externalism (Hurley 1998; 2010). On an ahistorical reading of active externalism, the concrete content of what is being processed appears secondary to the issue of the locus of cognitive activity, just as history is deemed secondary to present states of cognitive affairs.

Such a choice of priorities might be unfortunate, for several interrelated reasons – of which, to be sure, theories of extended cognition have become at least partly aware since that initial statement.[3] Nonetheless, it will be worthwhile to spell out why, in terms of a unified rationale, any serious theory of extended cognition will benefit from accounting for the history of the functions of environmental coupling.

First, when referring to the functions of some object or process, one will have to tell the difference between genuine functions and coincidental effects. That difference can be established only on the grounds of a history of how the object or process in question came to display the effects that it displays – and why it fails to do so under some circumstances. For artefacts, their design will provide part of the relevant criteria, whereas all relevant criteria will be historical in the case of natural systems.

Second, the content of cognitive activities – what in the environment they refer to and whether, why and how they succeed in doing so – is relevant to an explanation of the shape, function and extension of these activities, historically and at present, on more than one level. On the one hand, the concrete content of some cognitive process depends on the history of cognisers living in a concrete natural and social environment to which they relate. More fundamentally, on the other hand, the mechanisms for producing and acting upon these contents are rooted in histories of evolution and ontogenetic development, broadly construed.

Third, present interactions dynamically reshape the mode of extension in many cases. There is no ending to history. If the conditions under which some relation of cognitive coupling is realised are altered – from within or outside the organism – that change, unless ending in failure of the entire process, will provide some degree of variation that will result in a modified coupling relation further down the line. Variability in environmental conditions will be one key factor in shaping coupling relations.

The common rationale to unify this set of claims is an aetiological, that is historically based variety of functionalist argument. It departs from the classical Putnamian point of "machine-state functionalism", which is based on the assumption that one and the same function may be realised in a variety of different structures (Putnam 1975a). This claim of multiple realisability has been central to any argument for the possibility of AI, and it is equally important to the "parity principle" in the Extended Mind debates. It has its roots in Alan Turing's theory of computability (see p. 3–4 in Chapter 1), which says that any logical or mathematical operation that is formally solvable at all can be realised by elementary formal operations that may be implemented in a broad, in principle indefinite, variety of physical systems (1936). The claim of multiple realisability is also implied by observations on neuronal equipotentiality, according to which one cognitive or perceptual function can be realised by a variety of neuroanatomical structures (see, e.g. the discussion of Karl Lashley's work in Proust 1995).

While sharing with these types of functionalism the basic notion of a relation of underdetermination between functions and structural properties of some system, an aetiological account is interested in the concrete enabling and constraining conditions under which functions come to be established and realised. Functional aetiologies have their natural home in natural history and the Darwinian theory of evolution, where evolved analogies of function between phylogenetically remote species have been painstakingly distinguished from equally historically grounded, structural homologies that do not necessarily have any bearing on functions at all (see p. 5 in Chapter 1). Accordingly, the concrete histories of convergent and divergent functions within and between populations will matter to an aetiological account.

If cognitive traits have biological functions in the same way as other traits of an organism have biological functions, one will be entitled to analyse the mechanisms that realise these processes in the same functional-historical terms. Like any other trait, they will be subject to processes of variation and natural selection. In turn, the content of these processes can be analysed in analogous fashion, to the extent that a type of cognitive state of one or a number of related individuals is constituted by reproducible tokens that may succeed or fail to map onto some world affair, and hence be selected to accomplish that mapping. These are the basic assumptions of the teleosemantic or biofunctionalist paradigm in the philosophy of mind and language, as inaugurated by Ruth Millikan (1984).[4]

More precisely, a functional-historical account of some cognitive trait will recur to the contributions that the effects of that trait have made with sufficient frequency among its ancestral bearers to their rate of reproduction as compared with other members of the same population who did not possess that trait. By virtue of conferring a reproductive advantage over the course of several generations, the trait will acquire the *proper function* of producing these effects (for the full definition of proper functions, see Millikan (1984, 28 and Chapters 1–2 in general). The "ancestors" and "generations", rather than being organisms themselves, may also be component mechanisms, and they may also be reproducible forms that can be iteratively used by some individual or a collective thereof, such as artefacts or linguistic items. The individual tokens of some such reproducible item have proper functions derived from the functions of the mechanisms that produce them. It is the direct proper function of those mechanisms to generate the kind of effects in question if these effects, qua individually bearing certain relations to the environment with sufficient reliability, form part of the necessary conditions for the reproduction of the mechanisms themselves and of the systems that rely on their presence. In turn, the tokens of some reproducible item have the derived proper function of being adapted to some concrete world affair if and when one relevant condition for their reproduction lies in their co-occurrence with that sort of world affair and, in a relevant subset of cases, their role in directing an organism's behaviours towards it.

Nothing in this account rules out the possibility of mechanisms that are only partly based within an organism. In fact, Millikan (1993a, 179) herself considers the insight that "the organismic process has no skin" fundamental to the science of psychology.[5] Because all sorts of traits, cognitive and other, are characterised in terms of their historically acquired functions, and because that history incorporates whatever turns out to realise the function in question, the locus of the components involved in realising it is systematically irrelevant, under a two-part proviso: the organismic traits involved are to be coupled with persistent or reproducible features of their environment with a degree of reliability sufficient for meeting the reproduction criterion, where that coupling occurs in accordance with a uniform explanation that accounts for the "normal mechanisms" of coupling. Hence, although the content externalism defended in the aetiological account does not entail a vehicle externalism, it is methodically impartial to the locus of the vehicles involved.

As to the concrete benefits of an aetiological perspective for extended functionalism, accounting for the historical nature of functions will, first, provide an argument against the "coupling/constitution" objection brought forward by Fred Adams and Kenneth Aizawa (2001): in contrast to their objection, instances of accidental coupling between an organism and some entities in his environment do not count as coupling proper. Nor will legitimate instances of coupling endow the individual entities thus coupled with the characteristics of the entire coupled system. These entities play constitutive, historically established roles in the establishment or accomplishment of the functions of the entire system, where these roles may widely diverge. I will discuss this point in more detail in the next section.

Second, and for related reasons, accounting for the historical nature of functions will provide an argument against the "fleeting vs. persistence" criticism that has been levelled against extended cognition, first and foremost, by Rupert (2010). The suspicion here is that environmental extensions, unlike internally based cognitive capacities, are not persistently coupled with the organism. If some such extension is accidentally detached from the cognising organism, it seems, the cognitive system comes apart, only to be re-instantiated if and when that extension is successfully recovered. Clark's reply to this charge was to emphasise the conditions of persistence and reliability ("glue and trust"), which supposedly rule out overly ephemeral coupling relations (Clark 2010a, 83f). If the coupling in question has a historically established function, the issue of temporary non-occurrence or non-performance of that coupling relation, rather than being resolved by means of such auxiliary conditions, is actually *dis*solved: it does not matter if the function in question, for temporary loss of one component, turns out not to be performed in many instances, as long as the coupling occurs with a degree of reliability and according to an unequivocal regularity so as to become part of the explanation for the reproduction of the organic and, in some cases, the environmental components involved. To use a non-cognitive case of extension for comparison (see Menary 2010b, 14): it does not matter if a spider loses her web once in a while, or perhaps even very often, as long as the however intermittent production and use of webs provides part of the explanation for the presence of web-making mechanisms in the spider population.

Third, and most profoundly perhaps, a historical account of functions will counter Mark Sprevak's critique of classical machine-state functionalism (2009) which targets the "Martian intuition" that an identical or analogous kind of function can be realised by physically or biologically very differently constituted systems. On that intuition, which is basically the Lewisian one (1980), if we can specify a set of functions for certain traits and behaviours in human beings *and* Martians, we should be able in principle to identify which concrete structures within or around the human being or Martian will play what sort of causal role in accomplishing those functions. However, Sprevak objects, one could easily imagine the conditions of persistence and reliability of the respective structures and the coupling relations involved to be violated in manifold ways, and hence fail to meet the specifications laid out by extended functionalism. Playing on the Martian intuition in this way will probably kill the Martian – but, if we follow the historical account of functions, it will not kill the intuition. Any coupling with whatever structure that were as unreliable or otherwise insufficient or even freakish in nature as described by Sprevak would militate against the imagined organism's reproductive chances, and against the factual (rather than conceptual) possibility of his existence. Conversely, Sprevak's Martian may be imagined to be in command of capabilities and enter into environmental coupling relations so unfamiliar that we could not find any remote analogue in the specifications of human cognition. But then, the entire functional analogy will not hold. After all, there might be hyper-cognitive functions to Martians that human beings are insufficiently equipped to imagine in the first place.

Clark and Chalmers's reservations against history diagnosed here become explicit when they introduce a set of criteria that is supposed to be necessary and sufficient for the function of internal or external resources as components of cognitive processes (1998, 17; see also Clark 2010b):

(1) The resource should be of reliable availability, and should be used on a frequent basis.
(2) The information provided by that resource should be directly available.

(3) This information should be endorsed automatically, that is without requiring further reflection on its reliability.
(4) Present endorsement of this information should be based on conscious endorsement in the past.

The authors add a footnote to this list in which the implication of the constancy (3) and the past endorsement (4) criteria that history is co-constitutive of cognitive processes is countered by the suggestion that the past endorsement criterion might be dropped and the constancy criterion be given "a purely dispositional reading": one is found simply to endorse the information automatically; it is not asked why. Although this might be a sensible tactical manoeuvre in terms of catering for the broadest possible range of extensions of the mind, relativising the latter two criteria amounts to skirting an inquiry into the normal mechanisms of cognitive coupling. The past endorsement criterion in particular would assimilate the argument to a teleofunctional one: the decisive condition for past endorsement being causally relevant to present endorsement is that the former has been successful, although it might also have been conscious. If endorsement had not been successful, although it might still have been conscious, there would be no endorsement at present.

Beyond tactical considerations, one possible systematic reason why the proponents of the Extended Mind Hypothesis have been reluctant to adopt a historical account of cognitive functions is detectable in Wheeler and Clark (2008) – a contribution that actually acknowledges arguments for the rooting of cognitive extensions in evolutionary history: the authors match the dyad of extended/embodied cognition and cultural evolution hypotheses against the tenets of evolutionary psychology. The latter is charged with a view of evolution that takes psychological and physiological adaptations alike to be temporally remotely anchored solutions to a predetermined set of problems posed by the environment, where these solutions will be uniformly distributed over a species while being expressly domain specific. Moreover, Wheeler and Clark continue, the possibility of organisms modifying their environments to fit their specific needs is not considered by evolutionary psychologists. The basic concern here appears to be that an anchoring of cognitive extensions in past states of affairs that are beyond present cognisers' reach – especially if the past endorsement criterion concerns the *evolutionary* past – would undermine the notion of a two-way interaction that is central to their earlier cited coupled system claim. Adaptations and the conditions to which they respond might be more malleable and more specific to sub-populations in the here-and-now than evolutionary psychology appears to consider.

In order to get a clearer view of what is at issue here, a brief comparative look at Richard Dawkins's theory of the extended phenotype (1999) and other, more constructionist biological cousins of extended cognition, as they have been presented in Chapter 5, might help: Lewontin's environmental constructivism, niche construction theories and developmental systems theory. Prima facie, the parallels between extended phenotypes and extended cognition seem striking: Dawkins not only highlights the importance of an organism's interaction with the environment but also promotes the notion of biological traits extending into, or incorporating features of, the environment. What moves into focus in his gene-centric view are replicating gene sequences within a population and their interactions with their intra- or extra-somatic environment. Dawkins argues that the organism might not always be the key unit to consider when explaining patterns of genetic replication. Instead, he suggests, an organism's interaction with objects or other organisms within his environment will have to be counted into the explanation of whether and how a gene succeeds in replicating.

However, Dawkins's alternative, explicitly metaphorical proposal re-instantiates the notion of a unitary, individual and individualistic agent who is clearly distinguished from his external environment, albeit with a substitution of the referent of "individual". In many relevant cases, the manifest aims of individual organisms will end up subordinated to their genes' hypothetical interests. Moreover, the systematic point of adopting this view within the context of evolutionary theorising is to highlight a possible, and perhaps the most relevant, unit of natural selection – as the main force in evolution and the source of any goal-directedness or progress therein. The effects that the organism's own traits and behaviours might have on the course of evolutionary events will not figure prominently in such an account.

Wilson and Clark (2009) expressly endorse the continuum of constructionist, anti-adaptationist theories in biology that takes issue with these implications of Dawkinsian view, as those theories might bear more profound analogies to some of the central claims of extended cognition.[6] Their critique of Dawkinsian adaptationism can be unpacked into three positively stated leitmotifs: a principled openness of both evolutionary and developmental pathways, an emphasis on supra-individual factors as explanantia, rather than explananda, for the respective theories, and the acknowledgement of the interactive nature of the organism-environment relationship. On a first level, the notion of principled openness is expressed in the marked abstention from postulating some unifying force that would govern phylogeny and ontogeny. Natural selection counts as just one among other relevant factors in evolution, and it works on various levels, from genes to populations. Genes are just one among other relevant factors in shaping organisms and, possibly, transmitting information between generations of a population. On a second level, there is no reason to assume that there is such a thing as evolutionary progress or adaptive optima. What might seem like a perfect adaptation now may turn out to be a disadvantageous solution later, and there is no way of knowing in advance. Ecological change might always upset historically entrenched states of affairs. Current evolutionary states of affairs are historically contingent facts, providing no guidance as to how things should stand or where they will go next. On a third level, ecological change, both to the better and to the worse, might be brought about by the organisms themselves, in intended and unintended ways.

The constitution of cognitive extensions

What do the observations in the preceding section tell us about the Extended Mind Hypothesis? After all, all parallels to the debates within evolutionary theory were identified only in retrospect. Nor did the Extended Mind Hypothesis, as originally conceived, relate to teleofunctional theories. Still, arguments from either tradition might help to elucidate, and possibly achieve, Clark and Chalmers's explanatory aims, on two levels: the epistemological status of the hypothesis (discussed under the short form *Int*) and the question of the nature of extensions (discussed under the short form *Ext*).

The status of epistemic claims

Although the biological theories considered here figure as predictive theories for the most part, developmental systems theory and the theory of extended phenotypes deliberately assume a different (separate or additional) role, namely as challenges to seemingly commonsensical perspectives on their subject matter. As such, they might ultimately but will not necessarily result in predictive theories.[7] This dual notion of theories may serve as a template for a strong and a weak interpretation of the Extended Mind Hypothesis respectively: if, at least in important subclasses of cases, entities and processes in the environment are indispensable for the accomplishment or presence of some cognitive function, the expectation will be justified that the advocate of extended cognition shall identify the necessary conditions for those objects and processes to become part of that cognitive function while remaining impartial to what common sense might tell us. Some of Clark's own works (in particular 2007) as well as some complementarity-based accounts (most notably Kiverstein and Farina 2011; Sutton 2010) aim in this direction, in looking for real-world cases where the assumption of environmental extendedness of cognitive processes accomplishes tangible explanatory tasks. If, however, it is a mere matter of perspective to view some objects in the environment as components of cognitive processes, the claim is considerably weaker: it amounts to a suggestion to temporarily suspend common-sense beliefs in order to help us to a better understanding of some aspects of cognition.

The critique in Adams and Aizawa (2001), although not intending to capture this point, is diagnostic of what is at issue here. The authors explicitly mount a defence of common sense, presuming that common-sense views, *qua* being commonsensical, provide good guidance to cognitive inquiries. If, as Adams and Aizawa argue in alignment with a long-standing intellectual tradition (which includes, among many others, Karl Popper 1959 and W.V.O. Quine 1957), science is a methodical refinement and extension of common-sense reasoning, mobilising it against overambitious flights of imagination will be useful in itself. As, however, many scientific findings happen to defy common-sense beliefs, and as, arguably, the advancement of science sometimes even requires transgressions of the boundaries of common sense (as has been prominently argued by Albert Einstein 1918, along with other physicists whose theories indeed defy common sense, as well as by philosophers as different as Ernest Nagel 1961 and Paul Feyerabend 1975), the force of this argument will be limited to begin with. If there are defensible reasons for adopting a counterintuitive image of how the world stands, and if evidence collected under the guidance of that image turns out to advance inquiry, the transgression will be vindicated. It will be difficult though to determine in advance whether that vindication is forthcoming. In terms of an epistemological economy, the Extended Mind Hypothesis incurs the risk of investing into a transgression of common sense whose returns are much less than certain.

The nature of extensions

One common denominator of all non-adaptationist theories discussed earlier is that there will be no way of determining the goals and accomplishments of an organism without in some way considering his coupling with the environment, where this coupling will have the nature of a two-way interaction rather than the organism either manipulating or being determined by the environment. Certain entities and conditions in the environment are relevant to the organism in such a way that he could not exist without them and that they would not exist without him, as the very set of relevant conditions that they are. On this set of views, the coupling between organism and environment plays, in a non-Godfrey-Smithean sense, a *constitutive* role in either of two ways: without that coupling, a trait or behaviour would not be possible at all, or it would have to rely on different means.

Organisms with a certain degree of adaptive plasticity, that is the ability to respond in various and partly innovative ways to a given set of problems, can be found to alternate between using internal means and one or a variety of external objects or tools for the same purpose, each involving different organic and behavioural resources and, possibly, differential effects on reproductive chances. For example geographically separate populations of wild chimpanzees (*Pan troglodytes*) have been observed to vary in using vs. not using tools and in the kinds of tools they use for a certain purpose, in spite of identical biological constitution and ecological similarities between the respective populations' habitats (Boesch 1993; Boesch and Boesch 1990; Matsuzawa 2001, Chapter 1). It has been suggested that cultural transmission of tool-using abilities is involved here.[8] Although there is an analogy of function in these cases, both the entities involved and the mechanisms by which they are integrated in pursuit of that function will be at variance. The constituents of the external and internal variants of the processes cannot be expected to be mutually substitutable. This is the *minimal* sense in which extensions may become constitutive to the accomplishment of some biological function. They will be called "constitutive$_w$" where required for disambiguation.

The situation will be different in cases where there are objects and artefacts in the environment that enter into some coupling without which a particular, and possibly vital, accomplishment for the organism could not be attained *at all*. Citing a famous example, without felling trees and constructing a dam from timber, stones and mud, so as to impede the flow of a creek and create a water reservoir, a beaver would neither be in the position to secure himself a sufficiently large territory for foraging nor be able to find a place for building lodges sufficiently protected from terrestrial predators (Dawkins 1999, 200, 209). There is no 'internal' alternative for the beaver to building a dam. He does so because this is the only way available to him to secure the presence of some of the conditions on which he vitally depends. This is the *maximal* sense in which extensions may become constitutive to some biological function: extendedness will become essential to an explanation of the presence, rather than the mode of accomplishment, of some core functions – and, by implication, of the nature of the organisms to which those functions pertain. Cases of this kind will be labelled "constitutive$_S$".

The constitutive$_S$ and constitutive$_W$ types of case are clearly distinct from an interpretation of cognitive extensions in which they appear as *instruments* of cognition: if the cognitive function in question could be equally performed by purely internal means, their role would be similar to that of an artefact that produces a behavioural output identical to that of some type of human action, and that does so on the basis of at least similar mechanisms. Internal and external elements would be mutually substitutable without any notable effects on the performance, and the mode of performance, of the cognitive function. If (and only if) extensions of the mind were supposed to be understood in such a straightforwardly instrumentalist fashion – which, to be sure, they are not – a central point of Adams and Aizawa's critique of extended cognition (2001) would become tenable. The authors maintain that the causal processes involved in the use of cognitive extensions would have to be identical to those involved in internal or "intra-cranial" cognitive processes in order to count as functionally equivalent. Because the causal processes, as a matter of empirical fact, will be at variance, functional equivalence is not accomplished. If, however, the extensions Clark and Chalmers have in mind are not instrumental in nature but constitutive$_W$ to cognitive processes, the mechanisms involved in extended vs. internally based cognitive processes can be freely acknowledged to be at variance. Adams and Aizawa hence appear to miss the entire point of functional analogy, which is precisely *not* determined on the grounds of a homology between the structures that realise some function but on the grounds of what purpose, in terms of selected effects, the respective structures serve.

To sum up

We are presented with four different possible interpretations of the Extended Mind Hypothesis, on two partly interdependent levels:

(Ext$_C$) The constitutiveness claim: entities in the environment are constitutive$_W$ or constitutive$_S$ to the function of some cognitive trait.

(Ext$_I$) The instrumentalist claim: entities in the environment are instruments that are mutually substitutable with internally based cognitive traits.

(Int$_S$) The strong interpretation: the Extended Mind Hypothesis shall identify part of the necessary conditions for the presence and function of the cognitive traits in question.

(Int$_W$) The weak interpretation: the Extended Mind Hypothesis suggests some conditions under which cognitive processes are best viewed as including entities in the environment.

When introducing their hypothesis, what Clark and Chalmers appear to do is this: they present a set of theoretical claims that speak for an interpretation under which objects and conditions in the environment are part of the necessary conditions for the presence and function of some cognitive traits. Accordingly, the strong interpretation (Int_S) would seem in place here. However, these claims are quite casually placed alongside remarks on how a view of cognition as extended into the environment allows for a more elegant, unified and simplified explanation of cognitive processes (see Clark and Chalmers 1998, 10, 14). There is no reference in these places to explanatory problems that would *require* a view of cognition as being thus extended. Accordingly, the weak interpretation (Int_W) would seem in place here.

There are two adjacent claims on the first pages of Clark and Chalmers's essay where that ambiguity becomes manifest:

> If, as we confront some task, a part of the world functions as a process which, *were it done in the head*, we would have no hesitation in recognizing as part of the cognitive process, then that part of the world is (so we claim) part of the cognitive process.
>
> (Clark and Chalmers 1998, 8, emphasis in original)

After having thus introduced the parity principle, the authors present the core of the hypothesis, the coupled system claim, and add:

> All the components in the system play an active causal role, and they jointly govern behaviour in the same sort of way that cognition usually does. If we remove the external component the system's behavioural competence will drop, just as it would if we removed part of its brain. [. . .] The external features in a coupled system play an ineliminable role – if we retain internal structure but change the external features, behaviour may change completely. The external features here are just as causally relevant as typical internal features of the brain.
>
> (Clark and Chalmers 1998, 8f)

If, taking the first claim at face value, a part of the world worked as a process that could, in practice or in principle, also occur in the head, the 'external' and the 'internal' versions of that process would be equivalent, hence interchangeable without loss, and arguably without a difference beyond the physical location of that process. If, however, according to the second claim, we took some environmentally coupled cognitive process and removed the external component, we would *not* be left with an intact set of internal operations within the person accomplishing the cognitive task – only minus that component. Even less would we be left with a set of internal operations identical to the case of accomplishing the same task by internal means. As Clark and Chalmers implicate in the second passage, very different kinds of mental operations are required for doing a calculation in one's head and for doing it with a pocket calculator, or for finding one's way by visual recognition of a scene or by consulting a map or a notebook.

Clark and Chalmers's choice of examples is not precisely helpful in keeping these points apart. The examples they extensively discuss in order to substantiate their hypothesis – which are the examples that also dominate the Extended Mind debates – concern devices that we are suggested to consider as substitutes for, or supplements to, internally based cognitive processes. Their notebooks, keyboards and pocket calculators are artefacts that can be used for operations that could be, at least prima facie, equally accomplished "in the head", for example making a calculation or remembering an appointment or the directions to a certain place.

This choice of examples not only seems to suggest Int_W but also, however inadvertently, creates the impression that the extensions in question might be mere instruments of cognition – which is a logical *non sequitur*: although Int_S stands in a relation of mutual implication with the constitutiveness claim (Ext_C), the instrumentalist claim (Ext_I) and Int_W are logically independent. The claim that entities in the environment *might* be viewed as parts of cognitive processes does allow for the possibility but does not necessarily imply that these entities will work as instruments of a cognitive process, nor does the reverse hold. At the same instance, one will be entitled to accept that Ext_C and Ext_I might hold at the same time, albeit in different, non-intersecting domains: some extensions might play a constitutive role, and thus imply Int_S for the respective class of cases, whereas others might count as mere instruments of cognition in another class of cases. Still, Int_S will be vindicated, because there *is* a class of cases by which it is implied. On this analysis, the Ext_C type of cases provides much better support for the Extended Mind Hypothesis, and it does so for its stronger version.

In conjunction with the aetiological argument, these distinctions will contribute to a reasonably detailed classification of relations of extendedness. Any claim for a functional parity between the internal and external mechanisms involved will have to be cashed out in some not necessarily directly biological reproductive advantage determinable for each part of the equation.

If, first, I were able to substitute an environmentally coupled for an internal process at will, where that substitution made no difference to the achievement of my goals, in terms of scope, effectiveness, efficiency or reliability, the functional equivalence involved would be of the weak Ext_I kind. We still would have to explain how the elements involved have come to be thus substitutable.

If, second, the members of some population become equally able to mobilise some novel environment-based mechanism in order to supplement an established cognitive function, so as to achieve an increase in scope, effectiveness, efficiency or reliability, functional equivalence of a non-trivial constitutive$_W$ kind will be reached, while adaptive fitness will actually be increased across the entire population. This is where some of the standard examples of cognitive extensions will fit in, namely computing devices that increase the amount and complexity of available information or the speed of retrieval.

If, third, members of some population become able to individually mobilise a novel environment-based mechanism for an established function, so as to outperform other members of that population to whom the mechanism remains unavailable, the condition of functional equivalence will hold with respect to the sameness of function but not with respect to adaptive fitness within the population. Such would be the case for technologies of cognitive enhancement that are available only to some members of a population.

If, fourth, I use an environmentally coupled process in order to compensate for an internally based one that I am not able to perform for some reason, and if that environmentally based substitute enables me to accomplish the same task with roughly the same degree of effectiveness, efficiency and reliability (see Clark's "glue and trust" criteriondiscussed earlier), functional equivalence will be reached on all levels by means of a mechanism that is clearly distinct from the impaired one. Such will be the case in what has been termed "sensory substitution" (see Farina 2013; Kiverstein et al. 2015): situations of sensory impairment might be mended by artefacts that compensate for the function that has been lost and that typically recruit an alternative sensory modality.

If, fifth, an environmentally coupled process allows for the establishment of novel functions, and hence is constitutive$_S$ to them, there will be no reference point for determining functional equivalence, so that the function in question will have to be identified in the first place. It is upward along this hierarchy of types of functional relations that one will find the most rewarding proving grounds for any claim of extendedness.

Constitutional matters

The distinction between constitutive$_W$ and constitutive$_S$ types of cases introduced in the previous section partly matches the respective foci of "first wave" parity-oriented views of extended cognition, which highlight functional equivalence between internal and environmentally coupled cognitive processes, and "second wave" complementarity views, which highlight the role of environmentally extended processes as co-constituents of some cognitive abilities.[9] Still, the mapping of these "waves" onto the types of constitutiveness introduced earlier is not strictly disjunctive, and some of the preceding points might help to partly integrate these arguments: where classical functionalism focused on analogies between input/output relations in different systems on a general level, and where the "microfunctionalism" defended by Wheeler (2010) focuses on more fine-grained analogies in functions of the structures that realise a superordinate function, a history of selected effects of the traits of the systems under consideration will help to explain the establishment of certain functions in the first place. At the same instance, it is a matter of course to an aetiological view of functions to be impartial towards somatic boundaries, provided that the environmental structures in question form part of the necessary conditions in an explanation of the presence of some function. It is equally a matter of course for an aetiological view, as it is for the developmental systems theorist, to accept that the roles played by the co-constituents of some function are more likely heterogeneous in kind than mutually substitutable.

Still, one might accept either or both types of constitutive roles of environmental factors for some of the adaptive functions of some organism's traits without necessarily accepting them as being coupled with the organism in the sense suggested by Clark and Chalmers. According to Kim Sterelny (2010), the Extended Mind Hypothesis overemphasises prima facie functional similarities between internal and artefact-involving cognitive processes while playing down the role of the agents who not only rely on environmental supports but also actively mobilise environmental resources, natural and other, so as to construct their own cognitive niches (a concept more richly developed in his 2003 book). Sterelny (2010) argues that functional couplings of the Extended Mind kind are limiting cases of the more general phenomenon of environmental scaffolding: the reliance of organisms on the presence of some environmental features or artefacts without which some function could not be accomplished. For example human beings have become reliant on the availability of cooked food or means of cooking food, as their metabolic apparatus has become insufficient for coping with exclusively raw food. Hence, cooking is an environmental scaffolding for the human metabolism, but it will be hardly illuminating to refer to the human metabolism as being extended into the environment.

With respect to constitutive$_W$ cases, all hopes for an artefact to pass muster as being part of an extended cognitive processes rather than a mere scaffolding rest on the degree of integration of the artefact into the accomplishment of a cognitive function, such as memorising facts or orienting oneself in space – which are functions that still can be accomplished to some extent by the unaided mind. However, genuinely material "anchors" for certain cognitive tasks may often be both more efficient and historically prior to analogously formed conceptual ones (Hutchins 2005). Still, there will always be room for debate as to whether some concrete coupling with some artefact is tight and integrated enough to warrant the bestowal of the "extended" predicate, or to make us, in Clark's words, "natural-born cyborgs" (2003). There is no steadfast ontological criterion to make that decision for us, although the evolutionary history of the mechanisms involved will provide some guidance.

Hence, the quest will be for evolved mechanisms for recruiting artefacts for the performance of cognitive functions. Clearly, it is not some concrete type of artefact (the notebook, the smartphone) that we have evolved to be coupled with. What can be credibly argued to have evolved is a general and highly adaptable ability to create and recruit such artefacts – from simple drawings to computing gadgets. In Millikan's terms (see p. 113), it is the direct proper function of the organism-based mechanisms involved to recruit a variety of artefacts for the accomplishment of a cognitive task, and it is the direct proper function of any concrete type of artefact involved to make its contribution to the accomplishment of that task, in cooperation with the organism-based mechanisms. If, though, one takes the lessons to be learned from developmental systems theory seriously, the practice of using artefacts should be expected to feed back into the organisation of the brain, and might do so beyond childhood development (Farina 2016).

It will be more difficult to identify the nature and functions of instances of environmental coupling that play a constitutive$_S$ role, as the respective cognitive ability supposedly would not exist without the environmental counterpart. It will even be difficult to identify the ability in question as being properly *cognitive*. If one follows Adams and Aizawa (2001) in defining cognitive processes *ab initio* as a subset of the processes within the human nervous system, one has a criterion for what counts as cognitive that rules out "extra-cranial" processes by default, albeit a question-begging one. After all, there are many non-cognitive neuronal processes. It is the additional criterion of the involvement of "non-derived content" recruited by the authors that shall suffice to identify the subset of intracranial processes that are properly cognitive. In introducing that second criterion, the authors follow an argument in John Searle's critique of AI (1980): although linguistic items or other artefacts have the meanings they have and serve the functions they do because they are endowed, by stipulation or convention, with those meanings and functions, our thoughts have their meanings and serve their functions in at least partial independence from any such external fiat. The nature of the human mind bearing those thoughts is supposedly at least partly sufficient for the presence of that content. That nature is not further explicated though.

On the aetiological view, in contrast, both artefacts of all sorts and our thoughts have the meanings and functions they have because individual artefacts and mental tokens have been reproduced by some mechanism with sufficient frequency because they mapped onto some world affair, in terms of being connected to an appropriate set of behaviours towards it, reliably enough and according to a well-defined set of regularities. On this view, cognition basically is a specific activity of relating to one's environment, in which such mapping relations are used to produce adaptive behaviours. No principled difference exists between artefactual and internal mechanisms that accomplish this feat, namely producing the appropriate tokens on the appropriate occasions, where the conditions of appropriateness essentially depend on what in the environment the organism relies on. As we will see in more detail in the next section, the human faculty of language is the primary candidate for a properly cognitive trait that may both count as environmentally extended *and* fit the constitutive$_S$ condition, and hence militates against either of Adams and Aizawa's criteria: its function and contents depend on the operation of mechanisms that partly operate outside the human brain or body.

Entities and processes in the environment may contribute to cognitive abilities in constitutive$_S$ fashion in either of two ways, both of which include evolved mechanisms of coupling: on the one hand, they might be part of the necessary conditions for the realisation of the direct proper functions of a mechanism that is otherwise based within the organism. The entities and processes in the environment need not be reproducible items with their own direct proper functions, but their presence and use will be essential to the realisation of the functions of the organism-based production mechanisms. On the other hand, there might be mechanisms for the production of tokens of some type that are based in the environment while being coordinated with organism-based mechanisms, so as to jointly help him to produce behaviours appropriate to a given world affair in a given situation. It is in the latter, more complex type of case where reproducible artefacts will come to play a central role.

The art of coupling, basic and advanced

Clark and Chalmers only briefly and tentatively refer to relations that would count as constitutive$_S$ on the present account (1998, 11). However, these references are part of a brief evolutionary excursus and include two suggestions that match the distinction introduced at the end of the preceding section: the evolution of vision as exploiting the structure of local environments, and the evolution of language as a key structuring feature of human interaction.[10] Although Clark and Chalmers proceed to using language as their paradigm extension of the mind, it might be worthwhile to also pay attention to more basic modes of environmental coupling. The function of vision and language is certainly not one of instruments that could be used or discarded by human beings at will. In the context of human evolution, both are indispensable components of, and hence constitutive$_S$ to, human cognition. If an individual is deprived of either faculty, his or her cognitive functions will be impaired to some extent, and some functions could not be accomplished at all.

Still, there is a difference between vision and language, with their respective modes of environmental coupling, in terms of the depth of their evolutionary roots and the levels of cognition they feed into. Vision is not only the older and more common phenomenon in the animal kingdom. It can also be plausibly argued to be more directly tied to objects and structures in the environment, serving as a prerequisite for spatial orientation and object recognition for most animals, including humans. Language, in contrast, serves the coupling between human agents in the first place and presupposes a number of pre-linguistic cognitive capabilities, including mechanisms of orientation and object recognition. Moreover, whereas the environmental coupling in visual perception, in terms of necessary constituents of the visual process, occurs between mechanisms of perception and invariant properties of objects in the environment, language will rightfully count as a production mechanism in its own right.

On this view, the cognitively most foundational set of external entities to be coupled with is to be found in natural objects or processes in the environment. According to the ecological view of visual perception, as developed by James Jerome Gibson (1979) and discussed in Chapter 3, features of an organism's natural environment will provide the information to guide his activities, and will be available to him reliably, directly, implicitly and automatically, fulfilling all four of Clark and Chalmers's earlier cited criteria. Visual perception, as we saw, crucially relies on the perceiver's interaction with his environment and the objects he encounters therein. On the ecological view, inference or inner representations are largely irrelevant to perception, whereas none of these higher-order cognitive accomplishments could emerge without the prior establishment of mechanisms of perceptual coupling. Alternatively, between such elementary direct guidance by perception and fully fledged internal representations, one might find representations that detect patterns in one's environment and direct appropriate behaviours towards them at the same instance. These elementary representations have been introduced as "pushmi-pullyu representations" by Millikan (1995), and they figure as "action-oriented representations" in Clark (1997; see also Clark and Toribio 1994).

This image of directness has to be extended when moving from perception to higher-order kinds of cognition. On the one hand, some of the guidance that has hitherto been directly provided by perceptual couplings with the environment is now delegated to internal resources, some but not all of which will assume the shape of explicit representations. Thus, the next remarkable development, again in phylogenetic terms, does not as much lie in an extension of cognition into the environment as it lies in the internalisation of some of the guidance that would otherwise be directly provided by the environment. Externalisation reappears on the level of artefacts that are used for cognitive functions in such a way as to re-instantiate environmental guidance on a higher level.

On the other hand, the purposeful creation and appropriation of cognitive resources fosters their (partial) detachment from a present, imminent or otherwise predetermined practical use. In turn, however, the use of these artificial resources, once it has been properly learned, and if it functions smoothly, *normally* is quite as direct, implicit and automatic as in the more basic cases. One may pause to reflect on the information taken up from the environment, and one may then either resort to other sources of information or store away that information for later use. However, doing so remains a special case mostly reserved for instances of interference or error, and for (re-) assessing the goals of one's actions. In a familiar and well-behaved environment, the natural and artificial features of that environment will equally provide direct guidance.

Language has the intriguing property of essentially involving both artefacts and internal mechanisms in order to perform its functions. It is at once rooted in evolution and in the history of artefact use. More specifically, it has been argued, most prominently by Terrence Deacon (1997), that language co-evolved with the human brain and its capability of symbolic reference. Deacon takes language to be a structure whose properties and reproduction within language communities are in a straightforwardly evolutionary way interdependent with the properties particular to the human brain. The complexity and the adaptive functions of the human mind have as one of the necessary conditions of their emergence and present functioning the development and use of linguistic structures. The intra- and extra-somatic mechanisms for the production of linguistic items are tightly integrated with each other. Neither mechanism would be present nor could it function in absence of the other. Individually, they contribute to shaping their counterpart and its functions, historically and at present. Jointly, they enable speakers to create concrete artefactual structures on which their further interaction relies. This claim is to be distinguished from the considerably broader view that human linguistic capacities are a product of natural selection on the one hand, where the possible role of linguistic artefacts in shaping these capacities is not further considered, and Chomskyan linguistics on the other, where linguistic structures proper are supposed to be at once innate and *not* a product of natural selection.[11]

Complementary to the hypothesis of language-brain co-evolution, there are arguments for a co-evolution of tool use in general and human cognitive capacities. It is suggested that there is some covariance between an animal's brain size and structure on the one hand and his skills in using or manufacturing objects to manipulate other objects in their environments on the other.[12] Although the claim that there is a direct and necessary correlation between tool use and general intelligence remains contested (for critical views see, e.g. McGrew 2013; Shumaker et al. 2011, Chapter 7; Teschke et al. 2013), theories of this kind may enjoy somewhat more substantial empirical support in terms of paleontological evidence than the language-brain co-evolution hypothesis: early human tools have been preserved and can be classified into stages of development that accord with stages of human evolution, and closer comparisons are possible between tool use in humans and other species than between human language and animal signals. Such empirical support, patchy and contestable as it may be, helps to keep co-evolutionary theories of tool use at a comfortable distance from anthropological speculations in philosophy such as the concept of "homo faber", as criticised by Max Scheler (1928) or the diagnosis in Arnold Gehlen (1940) of man as an "incomplete being".

These observations about language and tools as constitutive$_S$ elements of human cognition, however, do not imply an either/or decision between them: the most plausible suggestion is that both tool use and language co-evolved with the establishment of modes of cultural transmission and complex social behaviours in such a way that only the combination of all these elements, rather than merely one or a few of them, sufficiently explains the complexity and function of human cognitive traits. It might also be that not only *cognitive* artefacts contributed to the evolution of these traits.

The constitutiveness$_S$ of tool and language use in general to human cognitive capacities has to be distinguished from the creation and use of any concrete artefact as being constitutive$_W$ to the accomplishment of some function for its user. Only under the former but not the latter perspective, one can say that the function in question is established by the use of artefacts as one of its necessary preconditions, and only under the former but not the latter perspective, one can say that it is also sufficient for the anchoring of that function as a biological one. The use of any concrete artefact may ultimately make a contribution to the make-up of some cognitive trait, and, if the historical accounts referred to here are substantially right, the use of concrete objects has been part of establishing the cognitive traits that we now call our own, but when considered individually, the use of a concrete artefact will not provide sufficient grounds for an explanation of why our cognitive traits are what they are, and what functions they serve.

If the human faculty of language and linguistic structures in conjunction are correctly described as achieving a complex cognitive coupling between agents and their environments that would not be possible without them, and hence be constitutive$_S$ to it, they may actually and, at the same instance, be the cognitive resource with the highest potential for *de*coupling, as they allow for public *and* individual use in a multitude of partly unanticipated contexts, and as they allow for degrees of abstraction and detachment from actual matters of fact that would otherwise be hard to come by. Hence, contrary to one concluding observation in Clark and Chalmers's essay, the human mind, without language, would *not* be "much more akin to discrete Cartesian 'inner' minds, in which high level cognition relies largely on internal resources" (1998, 18). If we may call our evolutionary kin as witness, our minds would be coupled with their environments in considerably tighter and more direct fashion, in that there would be little exception to being directly guided in our actions by what we encounter there.

If constitutive modes of cognitive coupling carry the day as the paradigms of extensions of the mind, as I have sought do demonstrate, we might better conceive of the human and other minds as extended not by virtue of being equipped with a number of artificial add-ons that work as external instruments of internally based cognition. Instead, we should conceive of cognition as being extended over traits of the organism and properties of the environment, incorporating and integrating entities, processes and mechanisms on either side of the somatic boundaries. Both sides have evolved, grown and, in part, purposefully created to be so integrated. In conjunction, they will either constitute$_W$ alternative mechanisms for the accomplishment of some cognitive functions or constitute$_S$ some functions, as necessary conditions for their presence. In either case, but to different degrees, the establishment and performance of these functions depends both on evolved mechanisms and on the environment-shaping activities of the cognising organisms themselves.

Whether developmental systems theory or environmental constructivism or niche constructivism or teleofunctionalism or theories of extended cognition are concerned, they appear to share a leitmotif introduced into modern Anglo-Saxon philosophy by John Dewey (especially in 1929).[13] In critical reaction to Herbert Spencer's self-styled Darwinian conception of the organism-environment relation and in accordance with a different lesson drawn from Darwin, Dewey considers the distinction between organism and environment artificial to begin with. Presuming that distinction to be ontologically grounded creates many of the problems that most accounts of the adaptive efforts of an organism keep struggling with. To live plainly *is* a particular way of being related to one's environment. It also is an interactive, environment-shaping relation, and it is continuous with being a cognitive system, in terms of the evolved function of cognition to help an organism cope with a complex and continuously changing environment. On this view, it will be difficult to find cognitive processes that do not incorporate features of the environment. At root, all cognition is extended, and only some of the historically most recent cognitive traits allow for a notable degree of decoupling. The mind is not extended *into* but *over* the environment.

Notes

1 For systematic mappings of that debate, see Menary (2010b) and Wilson and Clark (2009). Hurley (2010) embeds her account in a broader taxonomy of externalist positions in the philosophy of mind. A useful commented bibliography on extended cognition is provided by Kiverstein et al. (2013).

2 This concern is shared by the main critics of extended cognition, Adams and Aizawa (2001), Rupert (2004) and Sprevak (2009).

3 See, for example Wheeler and Clark (2008), where evolutionary and developmental considerations play a major role in an argument for dynamic interactions among genetic, organismic and cultural factors; or Wilson and Clark (2009), where an explicit and positive reference is made to evolutionary accounts of biological functions; or Rowlands (2004, 224–227), who refers to Millikan-style proper functions and their aetiological credentials.

4 Other statements of this paradigm include Godfrey-Smith (1994), Neander (1991a, 1991b), Papineau (1987) and Price (2001); see also the anthology by Macdonald and Papineau (2006).

5 Remarkably, this link between teleosemantics and extended cognition was first highlighted by one of the main critics of the latter: Rupert (2004, 401, n22).
6 For its emphasis on the nature of functions as, paradigmatically naturally, selected effects, and for its reliance on the positive selection *for* some trait, the teleofunctional account should be partly subsumable under the adaptationist rubric (these commitments are quite clearly stated in Millikan 2004, Chs. 1–2).
7 At the end of (2008), Clark asks for a "mental flip" explicitly styled after the one suggested by Dawkins (1999), whereas much of the preceding text is an attempt at providing concrete evidence for the explanatory value of extended cognition.
8 Among the few non-primate tool-using species, individuals of the New Caledonian crow (*Corvus moneduloides*) were observed to alternate between tool-guided and tool-less techniques of extracting food in the wild (Bluff et al. 2010). A recent example of allegedly innovative tool use and its social transmission in captive Goffin cockatoos (*Cacatua goffini*) is to be found in Auersperg et al. (2014). For an authoritative account of the varieties and complexities of animal tool use, see Shumaker et al. (2011).
9 The distinction between waves of arguments for extended cognition has been made by Sutton (2010). Defences of extended functionalism include Clark (2008) and Clark (2010b) and Wheeler (2010), while second wave views have been introduced by Menary (2010a, integrationism); Sutton (2010) and Kiverstein and Farina (2011, complementarity); and Rowlands (2010, amalgamationism). It was Clark (1997) himself who anticipated complementarity views.
10 For extended-mind-based accounts of language, see Clark (1998, 2006), and Wheeler (2004).
11 See Christiansen and Kirby (2003) for a collection of positions on the evolutionary roots of language and the possibility of brain/language co-evolution. For a classic, anti-Chomskyan account of the evolution of the human faculty of language, see Pinker and Bloom (1990), and for the position under attack here Chomsky (2006).
12 Studies focusing on the importance of tool use, typically in conjunction with other factors, have been assembled in Gibson and Ingold (1993), Matsuzawa (2001), and Sanz et al. (2013). Arguments for a primacy of cultural or "Machiavellian" intelligence over the use of linguistic and other artefacts are to be found in Byrne and Whiten (1988), Herrmann et al. (2007), Humphrey (1976), and Tomasello (2014); see also Sterelny (2003, Chapter 4).
13 For historical and systematic considerations of the Spencer-Dewey axis, see Godfrey-Smith (1996a; 1996b), Pearce (2014). Notably, Dawkins's extended phenotype theory may be considered playing the Spencerian part in this context.

8 The nature of cognitive artefacts

The term "cognitive artefacts" was introduced by Andy Clark (1998) – although it does not appear in Clark and Chalmers (1998). Kim Sterelny uses the roughly synonymous term of "epistemic artefacts" (2003; 2004) which, in turn, is taken up by Clark in (2010b). These terms are obviously supposed to be more accurate in definition than the notion of information technologies, which is both less well-defined and typically more narrowly applied to computing artefacts. The term "cognitive artefacts" is supposed to refer to that class of artefacts which, by whatever means, serve to support, augment or substitute the cognitive functions of their users, in an analogous way to tools and machines that support, augment or substitute human activities on a straightforwardly physical level. According to this analogy, information would be to cognitive artefacts what directed motion is to conventional machines (see the discussion in Chapter 1).

Although such an intuitive definition is not off the mark, it will raise questions concerning, first, what the nature of the information involved is, second, what cognitive functions are supposed to be thus served and, third, how, and hence by what kinds of artefacts, they are supposed to be thus served. I have dedicated much space to addressing the first two questions. In a nutshell, the answers are that the information involved is paradigmatically natural information, and that the cognitive functions involved are, at root, those of embodied perception and orientation in an environment, which are reliant on the active uptake of natural information. I will now direct my attention towards the nature of cognitive artefacts and the ways in which they provide information of the said kind and thereby serve cognitive functions of the said kind.

I will argue that there are two fundamental ways in which cognitive artefacts may convey information about a given subject matter: the information they provide is either convergent with, or isomorphic to, natural information that a perceiving subject could collect in his or her environment. This distinction, introduced towards the end of this chapter, will help to generate a more differentiated picture of the nature of cognitive artefacts than the conventional reality/virtuality distinction. It will be particularly relevant with respect to artefacts that integrate simulated elements into the perception of an environment, as in "mixed" or "augmented" realities. These examples and their implications will be discussed in more detail in Chapter 9. I will prepare this line of argumentation by first considering pictorial forms and their constitutive$_W$ role to the accomplishment of cognitive functions, and by then exploring the interplay between cognitive artefacts and informational environments on a more general level.

Being guided by pictures

Pictorial forms are a fairly elementary kind of cognitive artefact that may help us to a clearer view of the relation between natural and artificial cognitive resources. Pictorial forms function both on the individual level and on the collective level, in a partly analogous fashion to language. Without making a claim for a genuine theory of pictorial representation – or anything close to the paradigms set by Charles Sanders Peirce (1998) and Nelson Goodman (1976), I will understand pictures in a wide sense while keeping this concept distinct from the notion of images which, for including abstract mathematical mappings, is not necessarily perception related. A picture may be a photograph as well as a painting or drawing, but it may also be a visual presentation or representation of a more abstract and formal kind, such as a pictogram, a graph or a map. A picture may also be of the moving kind, such as a film or an animated simulation.

It would be difficult to provide solid empirical evidence for the plausible claim that one constituent$_S$ of the evolution of human cognitive capacities lies in the creation and use of pictures, and that it does so in a similar or even related way to what Terrence Deacon (1997) claims for language (as discussed towards the end of Chapter 7). If one sees the essential connection between language and cognitive capacities in that co-evolutionary process to lie in the evolution of symbolic reference, and if symbolic reference is to be understood in the Peircean sense here, as distinguished from iconic and indexical reference, pictorial and linguistic forms would *prima facie* appear as quite distinct affairs anyway:[1] symbolic reference is based on conventions and implicit norms of usage rather than qualitative similarities or causal connection respectively, whereas pictures would seem to be iconic rather than symbolic in referential character. Notably, however, pictorial forms, in their broad variety, encompass the entire range of the Peircean iconic, indexical and symbolic modes of reference, where even iconic reference may partly rely on conventions and implicit norms, and where symbolic reference is accomplished in a partly different and less articulated fashion than for language.

If the pictorial forms in question are supposedly realistic or, in Peircean terms, iconic representations of natural scenes or familiar objects, they can be presumed to capture and contain some of the natural information for perception that is, has been or hypothetically could have been present in the agents' environment. Even if the scenes or objects are idealised or fictitious, and even if they could not possibly exist, recognition of the type of thing depicted will be possible, as long as they are provided with some recognisable characteristics of real-world objects or scenes. One will recognise a fantasy picture of a band of unicorns roaming the grassy hills of a red-skied planet shortly before the setting of its two suns not least because the analogy to bands of horses roaming the grassy hills somewhere on Earth near sunset – things that one is likely to perceptually encounter in a real environment – is strong and clear enough.

Abstract paintings, graphs of statistical distributions or displays of the behaviours of exceedingly small, large or remote objects in a computer simulation, as visualisations of world affairs about whom no direct perceptual uptake of information would be feasible to begin with, and where no analogy of the aforementioned kind would be in reach, will provide different kinds of information: either natural information that is not accessible to direct perception, as in the case of graphs or other scientific visualisations, or information that should be considered genuinely artefactual, in terms of relating to concepts presumably devised by the item's author, as in the case of abstract works of art. Here, indexical and symbolic modes of reference, again in Peircean terms, will play a prominent role: elements of some picture will be supposed to indicate rather than resemble or otherwise exemplify qualities of their referent, or they come to refer to their subject matter by association or convention, and thus can only be interpreted within a conventionally based framework of reference.

As outlined in Chapter 3, the primary cognitive achievement of pictures might not be precisely that of being pictorial representations – understood literally, as *making present again*. James Jerome Gibson (1979, Chapter 15) argued that the notion of pictorial representation is misguiding, as the set of invariants captured by the picture would be confined to a few channels of transmission, and hence remain incomplete, and that the information they contain would be made available in detachment from the original context of presentation. This is not a purely negative claim. What a picture, at least one of the iconic variety, can achieve is to extract a limited subset of the information for visual perception that has been, would have been or could have been present to the observer, and to display it in a delimited optic array, which itself, for being so delimited (and not "ambient"), is ultimately placed in a non-pictorial environment – while remaining within the same framework of perception, namely visual perception, to which the same general principles apply. An iconic picture may be a partial record of perception, in that it preserves some of the invariants, hence some of the information, by means of which we perceive and act within our environments, and in that it presents that partial, delimited information in a context in which it is not naturally present.

Conversely, what an abstract work of art and a visualisation of a simulation in science can achieve is to present, in a delimited context, a specified set of information that would not, and perhaps could not, be available to visual perception and place it in a non-pictorial environment. Instead, it will first make available to visual perception what is depicted, as in a simulation, or it will supposedly alter the observer's perceptual and conceptual expectations, as many works of art are designed to do. In either case, explicit conventions or implicit understandings of how the pictorial form in question is supposed to refer to some world affair will have to be in place – from objects or processes that would be observable in principle but cannot be directly observed in practice, via theoretical entities to conceptual structures that are not necessarily supposed to bear any empirical reference. In some cases, visually pointing to the values of a variable or identifying and visualising patterns of relations within the target will be part of how the picture in question refers to its subject matter, and hence imply an indexical mode of reference. Both the indexical and the symbolic mode of reference may become part of an iconic picture, too (think of a photograph of a crowd at a rock concert, furnished with an arrow pointing to an individual and a caption "That's me at Sleater-Kinney's 1997 show in Cologne" – which the author of these lines happened to miss).

However, the question of modes of pictorial reference is in many respects independent of the Gibsonian point that the primary function of pictures will not be that of representation but that of providing a partial, delimited analogue of direct perception of an environment and the guidance of activities in that environment. In the latter respect, an abstract painting or a visualisation of a simulation might operate in a manner quite distinct from a photograph of someone or something I am acquainted with. However, the "mode of reference" variable only partly covaries with the "perceptual guidance" variable. Concerning the latter, the central question will be what guidance something as limited and de-contextualised as a picture could provide, and how.

A first approach to an answer will be to say that the possibility of pictorial guidance depends on a selection of invariants to be preserved by the picture that is suitable to the observer's needs. A detailed re-creation of what has been seen in the original perceptual situation cannot be the aim of a picture, but only the preservation of some of the information available in that situation. The observer may also be supposed to identify the mapping relations of pictures of quite abstract and formal kinds, and act upon them. The empirical rendering of a simulation is often designed with quite pragmatic criteria in mind: directing different observers' attention to what is deemed most relevant in the results. Indexicality may hence be combined with iconicity, but may also override it if what the observer is supposed to pick up so requires.

A second approach to an answer to the question of pictorial guidance will be to say that the information which is inevitably missing in the picture will be filled in by the observer: he or she may be related to the picture, or learn how to look at the picture, in ways that allow him or her to identify it as a picture of some concrete individual entity or event that he or she knows, or as exemplifying types of things or properties that he or she is familiar with. If the filling-in of the missing information is done properly, a picture may well guide the observer's behaviours in a way partly comparable to the presence of the original world affair. For more abstract picturing relations, as in simulations, the filling in will be of an equally more abstract kind. In order to make the unobservable variables observable, a preliminary concept of the structure and behaviour of the target has to be in place when first-hand experience, via perceptual encounters with the target, is either unavailable or insufficient.

A third approach to pictorial guidance will specify that one key element of filling in the information that is missing in the picture will be to know how to apply the relevant transformations of the mapping relations between the original view of a scene or an object and those involved in seeing the picture. On these grounds, Ruth Millikan (2004, Chapter 9) argues that, if we know what transformation rules are to be applied, we actually *see*, through a different medium but nonetheless directly, what is depicted. (Even language, on her account, allows for direct perception of the subject matter of some utterance.) It will be a relatively straightforward matter to keep track of the transformations of the mapping relations for cases such as the photographer's view of the scene from the mountain on which the picture was taken vs. my view of the scene as it is now pictured in the book on my table. Keeping track of the transformations will, however, be more difficult for more abstract ways of depicting world affairs, where there is no obvious way of anchoring the transformations of the mapping relation in such a thing as an original view of a scene. Besides some concept of the target, a concept of the required transformations will have to be in place.

Among these three conditions – suitable selection of information, availability of contextual information for filling in the missing information and knowledge about the required transformations of the mapping relations – the latter two might suggest that perception of pictorial forms is mediated through theoretical concepts and hence indirect. Although there undeniably is a level of conceptual mediation to this kind of perception, that fact neither provides a criterion of demarcation from direct visual perception of natural objects and scenes, nor does it imply an indirectness of perception of pictorial forms. If all three conditions are fulfilled, the information about the target that is contained in the picture will be preserved, and hence be perceivable to the observer. It is the informational limitations and de-contextualisation of pictorial forms that will affect all kinds of pictures, but these limitations as such do not make the reference of the pictures indirect. What will be required is the filling-in of contextual information and the application of transformations of the mapping relations, which amounts to conceptual mediations. Once, however, these mediations are firmly in place – which may require learning how to look at a certain picture or kind of picture – information uptake may become quite direct, so that one will be entitled to say: "Das Bild sagt mir also sich selbst" (Wittgenstein 1969, IX, 115).

However, what mediation implies is, first, that a picture, like a linguistic item but unlike an instance of misperceiving a natural object or scene, might guide us with respect to things that do not exist without necessarily losing its referential character, and without necessarily amounting to an artfully created illusion. A picture of a band of unicorns at double sunset on a distant world, although neither that world nor those animals may exist, will still preserve some of the information that would have been present in that scene if it existed. At the same instance, it will be clear to the observer that the animals, the hills on which they roam and the two suns are not constituents of his environment.

Second, there might be considerable variation between different observers in terms of available contextual information and knowledge of the required transformations of mapping relations, and there might be no obvious and unequivocal way of mending it: is the picture in question a photograph taken during my vacation for private use, is it a painting of a person or place that can be assumed to be known to a collective of people, or is it a map or pictogram designed for public use? Is the other observer of a visualisation of the hydrogen bomb core simulation versed enough in nuclear physics to see the importance of the change in the target at t_n?

Does he share the theoretical views that make this change relevant in explanatory terms? How can we make sure that a variety of observers relate to the same picture in the same way, so that its function as an external cognitive resource is secured? First and foremost, the information required for a common understanding – for a shared mode of filling in the missing information, as it were – may be presented in a variety of ways, with different degrees of articulation, standardisation and public availability. It may consist in a picture's placement within a series of pictures, or in different individuals' memories of the event depicted, or in an act of pointing towards what is depicted (if it is present in a shared environment) or in some commentary or description in natural language.

Arguably, the last of these ways of providing contextual information will be of the highest degree of articulation, standardisation and public availability – which, however, comes at a cost. In their "Extended Mind" paper, Clark and Chalmers justify the privileged status they assign to language as an extension of the mind by its function of achieving a cognitive coupling between individuals, so as to jointly meet individual and collective purposes of the cognitive or practical kind – rather than coupling one organism with aspects of his environment so as to meet a certain set of individual purposes. For linguistic items, their proper way of usage will be fixed and occasionally modified in the very course of public usage and, to some part, by explicit definitions. Moreover, the rules thus established can be explicated and reconstructed by using the very same cognitive resource. On the grounds of these properties, language may also be used to negotiate and fix the reference of pictorial forms, and to give their users some guidance towards the proper way of looking at them – although practical training of one's visual system and some degree of acquaintance with the object or kind of object or scene pictured will be no less important for this purpose.

Considering maps or graphs, the role of language in properly understanding them is quite obvious, as they are usually equipped with instructions on how to read them (e.g. captions or legends that reveal what type of symbol stands for what type of referent). Language is contained in them or attached to them. For other pictorial forms, no such obvious and, as it were, institutionalised role of language is available. One may give descriptions of some elements of the picture, telling the innocent observer what he or she ought to see there, or one may describe the context in which the picture was taken, so as to remedy the asymmetries in contextual information, and perhaps establish rules of looking at a certain picture or type of picture.

It may be argued that such asymmetries in contextual information pertain to the use of linguistic forms, too, but these asymmetries are easier to track in language, as they can be explicitly addressed within the same medium, which, at the same instance, is the medium with the highest degree of articulation that we have at our disposal. It would be much more difficult to comment on a picture with a picture, so as to disambiguate its content, and to do so within a publicly accessible, explicit framework of reference that would still remain pictorial. The cost of filling in the information missing in the picture is that the mode of presentation will be changed, in moving away from visual perception. Such mediation, although it helps the observer in seeing what is depicted and in doing something with, to or about it, gets in the way of a coupling that would be at least similarly direct as in visual perception of world affairs in a natural environment. If it is indeed a detour for reality to explain itself through language ("Für die Realität ist es doch ein Umweg, sich über die Sprache zu erklären", Wittgenstein 1969, IV, 114), it may be a necessary detour when it comes to fixing the content of a picture that does happen not to speak to me in the same way as the familiar environment in which I perceive and act would speak to me.

Pictorial forms, as I have tried to show, form an interesting example of a cognitive artefact that sits halfway between direct guidance by features of one's environment and the language-bound possibilities of decoupling, and thus for abstracting from matters of immediate practical relevance or factual existence. They share some of either's merits and difficulties. At the same instance, they might be the artificial cognitive resource that, despite the limitations discussed earlier, displays the closest analogy to perception of natural objects and scenes under natural conditions. Although it will remain difficult to prove that pictorial forms are constitutive$_S$ to cognitive functions to the same extent as, arguably, language or tools are, they certainly will play a constitutive$_W$ role in many contexts of interaction with and within human environments, and they, like language, will be implied in the use of other cognitive artefacts.

Cognitive artefacts and informational environments

Arguably, the distinction between constitutiveness$_S$ and constitutiveness$_W$ of artefacts to cognitive functions admits of degrees. Even if an artefact is not constitutive$_S$ in a strict sense, the effects of its use are likely to feed back into the constitution of cognitive traits and their functions. These effects will typically not be direct, in terms of becoming selectively relevant to the reproduction of a cognitive trait, but manifest themselves in modifications of their users' informational environments.

There are several ways in which cognitive artefacts may have such modifying effects on variables in informational environments – if these effects are persistent and consistent. An artefact may, first, induce perceptions of things that are not present in the environment, such as, for example, in the case of iconic pictures of objects, as discussed in the previous section. Second, it may induce perceptions of things that are present in the environment but not within the normal range of human perception, such as in a variety of sensor technologies. Third, objects in an environment might be arranged to appear different in certain respects from how they are, such as in Gibson's example of a glass-plate extension over a cliff, which may be stepped and walked upon but which does not provide a stepping-on affordance. Fourth, objects might be arranged to appear largely indistinguishable from natural, real-world objects within a simulated context. In all of these cases, part of the make-up of *R*'s informational environment will have been altered.

Hence, with respect to the effects on the environment, the difference between cognitive and other artefacts lies in the ways in which our environments are made to serve our purposes. Tools and machines are mainly used to materially modify conditions in the environment, in the way outlined by Peter Godfrey-Smith (1996a) as "causal construction" (see p. 87). In contrast, cognitive artefacts might be found to have only modest effects in terms of materially modifying conditions in the environment, but they will affect the information that is available in the environment and the way in which it is available, and hence contribute to users' "constitutive" relations to their environments.

Let us briefly recall that, although constitutive relations, as understood by Godfrey-Smith (and thus, for purposes of disambiguation, referred to as "GS-constitutive"), are always informationally based, the distinction between the construction and GS-constitution of features of the environment cuts across the distinction between physical and informational properties of the environment. After all, both constructive and GS-constitutive features might take part in furnishing information to the organism, or in placing additional physical interventions in the environment. Territorial markers are physical entities that serve an informational purpose – one that will not, or not fully, be deducible from the physical nature of the marker. In a similar vein, cognitive artefacts, being artefacts that have the function of supporting, augmenting and, in some cases, substituting their users' informational relations to their environments, incorporate both a constructive and a GS-constitutive aspect. However, the organism's informational needs and purposes will typically override the physical aspects.

On the argument on cognitive extensions in Chapter 7, the informational accomplishments of the artefacts under consideration here typically are not mere equivalents of, or functionally neutral additions to, natural human faculties of collecting and acting upon information. They are so for at least three related reasons:

(*e* 1) Cognitive artefacts introduce new mechanisms of collecting, structuring and presenting information.
(*e* 2) The operation of cognitive artefacts might, as a secondary effect, affect the purposes of their users.
(*e* 3) Cognitive artefacts may, purposefully or inadvertently, alter their users' informational environments.

If the introduction of a cognitive artefact *neither* changed the mechanisms of collecting information that I have at my disposal *nor* affected the goals that I could attain *nor* affected conditions in my informational environment, the use of the artefact in question would be perfectly identical to the use of the natural, internally based cognitive resources they supposedly supplement, augment or substitute. However, this is a theoretical possibility that has little chance of being borne out empirically. The three factors listed previously will individually or, more probably, jointly affect the collection and provision of information by cognitive artefacts, in analogous fashion to what artefacts in general effect in terms of providing new mechanisms and altering their users' purposes and environments.

To begin with, one is not likely to find many artefacts that accomplish a task in the same way as the unaided human organism will. A stick might qualify as an extension of the arm in a straightforward, almost trivial manner that fits Ext_I (see p. 119 in Chapter 7), but still, it might be used in a variety of ways that are not necessarily of the length-extension kind. At any rate, most tools and technological artefacts will bear a more complex relation to human behaviours. Returning to the Gunderson (1964) quotation in Chapter 1, the steam drill's success in digging railway tunnels did not prove that machines had muscles but only that there are other ways of digging railway tunnels than by flexing muscles in an appropriate way. When it comes to cognitive artefacts, there will be, as a matter of empirical fact, no such thing as an extension that is strictly instrumental in character, although some such artefact might be conceivable. We should expect mechanisms to be variant by default. Otherwise, we will be restricted to trivial cases, such as spectacles or contact lenses. (Even hearing aids might not be a straightforward Ext_I case, to the extent that their sound processing mechanisms differ from what happens in the human ear.)

As an implication of the variance in mechanisms involved, the use of artefacts is found to affect the purposes of human action with reliable frequency. Continuing on the earlier example, if one can build railway tunnels more affordably and more quickly using steam drills rather than mere manpower, the reach and use of railway traffic is likely to grow, which, in turn, may produce an array of partly unexpected effects on the purposes of railway travel. Such effects may expand the scope of one's activities, or they may turn out to undermine one's purposes, in the same fashion as the effects of an organism's or population's behaviours, according to the Lewontian view of constructing environments, may equally foster and undermine the fulfilment their biological purposes.

On the analysis of informational environments presented in Chapter 6, there are several possible ways in which technologies concerned with the provision and processing of information might affect informational environments. If an artefact provides me with information about conditions in my environment that are relevant to my cognitive purposes, and if that information were practically impossible or more difficult to obtain, less complete or less reliable without the use of that artefact, or if it adds new kinds of information to that environment, that artefact's operations, under the condition of its normal functioning, will have changed my informational environment in terms of enriching it or, to refer to Kim Sterelny's previously quoted terminology (2003), making it more "transparent". Conversely, if an artefact interferes with my activities of information pickup so as to make signals persistently *less* reliable or even unavailable, my informational environment will have changed, too, namely in terms of being impoverished or becoming more "opaque".

The positive vs. negative nature of a change in informational environments is not correlated with the purposeful vs. accidental nature of that change. On the understanding proposed here, modifications of informational environments, although they *might* be purposefully created, do not always end up being conducive to human cognitive purposes, even if deception has not been the aim. Conversely, although changes in informational environments *might* result in illusionary appearances, ambiguous perceptual situations or misperception of affordances, this does not mean that they are due to processes that involve deception or dysfunction. To recall, the change in question has to be, first, *persistent* in order to qualify as a change in an informational environment (see p. 104 in Chapter 6). The introduction of new affordances or modifications of existing affordances by whatever means needs to be well-structured, which means that a limited and clearly circumscribed set of the variables within *R*'s informational environment is subject to a non-transient change. Second, the change in question requires a coherent pattern of modification, and hence has to be *consistent* in order to be accommodating by some means for *R*. Whether the change opens up new informational or practical opportunities or increases informational uncertainty, it must allow him to methodically alter his responses towards the signals or towards what is being signalled.

To illustrate my threefold point in (*e* 1, 2, 3) with a contemporary example: if I have information on my precise location available at any time, using a mobile device that displays to me my GPS (Global Positioning System) coordinates projected into a map-like visualisation, I will be provided with information that is factually identical with the information that I would pick up from a map that I have learned to read, from a verbal description or from an environment with which I am familiar. In the latter two cases, however, I will not have all that information available at all times, or some of it may not be available at all. For example I have to fill in the information on my current location myself when using the map, which I might sometimes not be able to do. Or I must have at least partial acquaintance with some place when trying to get my directions only from looking at my surroundings.

This example might look a bit mundane, and it was deliberately chosen to do so, because even in such a comparably simple case, which involves a technology that is not intelligent on any AI-related account and whose use is presumably not ecologically or selectively relevant in a life-or-death manner but may become practically relevant in many respects and in many situations, all three conditions mentioned previously apply: the mechanisms involved will have changed, my purposes will have been affected and the availability and quality of the information on which I rely will have been modified. Think of a world in which almost everyone is almost always guided by GPS devices, a world where maps and signposts have been consigned to obsolescence as archaic, primitive and insufficient means of orientation, and where that device-based guidance is designed to feel so natural that the use of the technology is hardly noticed by its own users, as in Mark Weiser's vision of "ubiquitous computing" (1991) quoted in Chapter 1.

Concerning the issue of mechanisms of information uptake (*e* 1), different cues are used, and different skills are required for different modes of accomplishing the same task, namely orientation. I do not need language to find my way by purely environmentally bound visual orientation. Recognising a landmark might involve the use of language but could, in principle, be accomplished without it. I do need language to learn reading a map, but once familiar with it, guidance will be more direct than from a description in natural language, as my orientation remains operating on the level of visual cues. Asking someone for directions will be helpful if I get lost and cannot match what I see on the map with what I see around me. In the case of using the GPS device, both learning and using it will be less closely tied to the use of language than for maps. A well-designed device will help me to track the domain of the signals of my location quite directly, and to readily apply the appropriate transformations of mapping relations between what is displayed on the screen and what I see in my surroundings. Still, the cues I use will be different from natural, unaided visual orientation, and the requirements to my judgement as to whether I am still on course will be different from both natural visually based orientation and reading a map. Important topographical features might be misrendered by a GPS device or a map – which one may visually detect by comparing what is displayed with what is in the environment. GPS coordinates might be miscalculated by the device – which will be harder to detect, given that I normally do not and cannot track the process of calculation within the device.

Concerning the issue of modifying purposes (*e* 2), the reach of my activities has changed by allowing, under normal conditions, for rather effortless orientation in unknown surroundings. In principle, I could go anywhere, and perhaps remotely interact with other people who use similar devices, without knowing anything about the topographical features of the place or the presence of those people in advance. Using a device that provides all the information that I otherwise might not be able to fill in from observation and experience alone, I am likely to be more directly guided by such a device and remain more immersed in my other activities than in finding my way. I am likely to behave differently when moving to an unfamiliar place, in not consulting a map beforehand, or in not asking people for directions, and, arguably, in relying on whatever the display of my device displays to me, even if the system is in error.

Concerning the issue of changing informational environments (*e* 3), if the GPS device turns out to make my activities of spatial orientation persistently more complex and less reliable, either for providing ambiguous signals or for its signals, despite proper informational mappings, not being properly recognisable to me, I will have to resort to other available cues in order to accommodate for this situation, or else find my plans frustrated. If, however, the use of the device opens up new opportunities of acting in space and time, and if these opportunities depend on altering the cues I normally use, there will be a more positively connoted change of my informational environment. The GPS device might make spatial orientation easier and more reliable after all. Even in the latter case, some amount of accommodation will be required. I have to learn to use and rely on the altered cues.

The picture receives some additional complexity in the case of artefacts that are supposed to keep track of, and react to, my behaviours and include this information into what they display to me, such as in advanced driver assistance systems, a concrete example of which will be presented in Chapter 9. In systems of this kind, information is collected and made accessible to me about my behaviours or cognitive states. Am I about to fall asleep behind the driver's wheel? Am I drifting off the traffic lane? Such a device may also collect and display information about my own relation to my environment. Am I getting distracted by my child's behaviour just as a critical situation emerges ahead of me? Information on the traffic situation and my behaviours is integrated in order to raise my attention to the situation at hand.

In cases of this kind, my purposes are certainly affected in the most straightforward fashion. Moreover, my informational environment is altered, in terms of cues being added that are relevant to my further course of action but that would remain, by definition, as it were, out of my perceptual reach in the kind of situation at hand. I will have to learn to make use of those additional cues. A driver assistance system that is successful in improving road safety is supposed to change my and other drivers' ecological environment for the better, in terms of reduced risk of physical harm. The informational relations involved, including the information the system collects about my behaviours and cognitive states, are still part of domains of natural information though, as they could be tracked by a suitably placed real or hypothetical observer – but not the receiver immersed in the situation.

Convergence and isomorphism

In the two examples briefly discussed in the previous section, the information involved retains some key properties of natural information. Although there will be some variance in the mechanisms involved in information uptake and transmission, the relations between the purposefully produced signals and the world affairs onto which they are supposed to map remain largely continuous with the relations the receiver would detect when using natural cues in his environment.

More precisely, the subclass of cognitive artefacts exemplified by these devices is marked by their ability to provide real-time information for perception about some condition in the environment, and by providing it in the course of being used within that same environment, so as to guide their users' decisions and activities in a given situation. In this class of artefacts, the analogies to the use of natural information for perception will be as close as factually possible. Their primary purpose is to provide information that is integrated with natural information for perception in the natural environment in which they are used – a condition that is satisfied by artefacts as humble as signposts but developed to the fullest by Augmented Reality Technologies, which integrate virtual elements with a naturally perceived environment. I will discuss these latter technologies in more detail in Chapter 9.

This subclass of cognitive artefacts has to be distinguished from a second subclass, whose purpose is to provide information that relates to a largely or entirely closed, artificial environment which is supposed to be perceived and treated in analogy to the perception of a natural environment. Hence, the environment about which there is information will be created by or within the artefact. The domain of the information involved is closed and self-contained, and typically can be clearly identified as such, independent of the degree of realism involved. Such will be the case, to a limited extent, for pictorial forms that do not picture real-world objects and scenes, whereas the paradigm of this class of cognitive artefacts are Virtual Reality or simulator environments, such as flight simulators.

What can be claimed for the first subclass of cognitive artefacts but not for the second is that the domains of the signals produced by the artefact are *convergent* with the domains of natural information that is or could be available to the receiver in a given situation. The term "convergence" assumes a technical meaning in this context:

(DS-1) Convergence is achieved if, for each domain D_a of a type of signal $\mathbf{r}_a$ that is emitted by some artefact and observable for *R*, there is a domain D_n of another type of, ultimately natural, signal $\mathbf{r}_n$ that can be, or hypothetically could be, observed by *R* while both domains are supposed to be correlated with the same conditions *F* at the source *s*, and transformations thereof, by the same or corresponding regularities of informational mapping.

If information basically is the co-occurrence between a proximal and a distal world affair in which conditions at the distal end covary with conditions at the proximal end in accordance with a strict natural regularity (see NI-1, Chapter 4), and if some artefact is supposed to track the same conditions at the distal end that the unaided receiver seeks to keep track of, although possibly by different means, the condition will still hold that any transformation of conditions at the distal end shall be mapped by transformations of conditions at the proximal end. If the artefact, in order to track the distal condition, records the values of the *same* proximate variable as the human receiver by means of a sensor that is built differently from the respective human sense organ, the mapping regularities will be virtually identical for D_a and D_n, even if that sensor is more receptive or capable of recording a wider range of values than the human sense organ. If, however, the artefact uses a *different* proximal correlate of the distal affair in order to track the same conditions at the source that the human sense organ is in pursuit of, transformations in the respective proximate correlates will have to correspond in order to meet the condition of convergence. If conditions at the source change, either proximate variable will have to change in such a fashion as to not only ensure a mapping onto distal conditions individually but also provide a reliable or, for the purposes at hand, reliable-enough mapping between the transformations of the values of the natural and artefactual proximate variables.

For example if the pattern of visible radiation of some distal object were used as a cue of a certain condition of that object by the human observer while patterns of thermal radiation of that object were tracked by an artefact, the values of the variables recorded by the human observer and those recorded by the artefact are bound to correspond only for some objects. Although a source of electromagnetic radiation emitting a spectrum of waves quite similar to that of the sun, as filtered through the Earth's atmosphere, will preserve correspondence in this respect, anything emitting a different spectrum of electromagnetic waves may not. LED lights will be found to shine very bright without getting very warm. A piece of iron will get considerably hot in the process of being welded or forged well before emitting a visible glow. Hence, either the range of objects to be tracked has to be restricted to those where correspondence between visible and thermal radiation holds, thus ruling out LED lights and hot iron, or the condition of convergence will not be met, for want of the required correspondence.

Even if correspondence is achieved, the sets of domains D_a and D_n do not have to be fully coextensive, as some artefact may be trained on a subset or superset of the information that is available for direct perception. Moreover, convergence does *not* imply that the artificially created signal is a signal of the natural signal and, in this sense, second-order information. Such second-order information constitutes a specific subcase of informational relations that will obtain only if there is a device or a natural mechanism with the function of signalling the presence of natural signals. Language is a mechanism that is often used to accomplish that.

If the natural information in a given situation is of the intrinsic kind discussed in Chapter 3 (p. 45), that intrinsic character will typically also be preserved for the receiver, although it is conceivable that it may turn extrinsic. Where in the former case receiver-dependent changes in the character of the affordance – for example when body-scaled information is involved – will have to be mapped by whatever artefactual rendering that can be accomplished, that affordance will be replaced by a display of measurements of physical variables in the latter case. Convergence can be achieved either way, depending on the functions of the artefact in question and the purposes of the receiver. After all, he might have to rely on intrinsic information in some kind of situation (Is that item too bulky for me to carry?), and on extrinsic information in another (Is that cupboard too wide to fit into the niche in our dining room?).

The requirement of convergence will also hold for devices designed to aid or sharpen natural perception. Enlargement, amplification and other effects have to preserve the informational mappings their user would pick up if his sense organs were capable of the degree of discrimination available to the device. Similarly, the requirement of convergence will hold for any device designed to collect and display information to me that, given my biological constitution, I could not obtain even with an exceedingly discriminating perceptual apparatus. A Geiger counter is definitely supposed to track the domain of signals of nuclear radiation, which is natural information that might become relevant to me under certain circumstances, despite being inaccessible to the natural perceptual equipment with which I am endowed – hence the condition of *hypothetical* observability in the definition of convergence. This example also provides an additional illustration of the difference between ecological and informational environments: Human beings and most other species are reliant, in a strictly biological sense, on nuclear radiation in their surroundings constantly remaining below a certain threshold. This is an ecologically relevant condition, and hence will be part of the ecological environments of all those species. However, none of these species (and, to my knowledge, no extant or historical species on Earth) has any sensory equipment for picking up information on radiation levels. Hence, the uptake of signals of radiation levels is not normally part of their informational environments.

A markedly different kind of situation emerges when we consider a simulated environment: if it is designed with a realistic representation of some natural environment in mind, anyone using the simulation via an interface that allows for a high degree of naturalness of interaction will use the same visual cues as he or she would use in a real environment in order to orient himself or herself and pick up information on objects and events. However, there will be no such natural environment from which to pick up any information, but a mostly iconic representation of such an environment that serves as the information source. Under these conditions, the user will use a subset of an analogue of the cues that he or she would use in a natural environment, and he or she will pick up information on analogues of objects and events that he or she would encounter a natural environment.

A paradigm of such a simulated environment is a flight simulator, whose purpose is to give pilots a realistic impression of the experience of flying an aeroplane. Events within the flight simulation call for operations on the simulation's controls that are, in their effects on that simulation, analogous to the effects of the same operations in the flight that is being simulated. Hence, the simulator environment is equipped with a cockpit with displays and controls that are all tied to the simulated conditions in such a fashion as to produce a partial, but in the selected aspects credible or even near-perfect, replica of the experience of flying an aeroplane. Pushing the lever connected to the simulated elevator too far *will* result in a simulated nose-dive, which might be displayed to the pilot in some richness of detail, and to which he can react by further simulated manoeuvres that map onto their real-world counterparts analogously. Neither the physical or functional structure of an aeroplane will have to be reproduced for this purpose nor, of course, the physical effects of handling or mishandling an in-flight routine.

In simulated environments of this kind, the primary change involved does not concern the mechanisms of information uptake (*e* 1) but the creation of a self-contained and well-ordered domain of informational relations. In a simulated environment, there might be no information on concrete conditions at some individual source in a concrete target environment, that is the environment that is being simulated. Concerning the possible alteration of the informational environment (*e* 3), the situation is intriguing: on the one hand, the simulation does not present natural information to me, as the reference of the displays and controls is purely internal, directed towards and contained within the simulated environment. In this respect, there is no common reference point in concrete natural informational relations with respect to which there could be a change. Hence, the condition of convergence (DS-1) cannot be met. On the other hand, the pilot using the simulation will become part of an environment that, although being thus self-contained, is supposed to faithfully represent a selection of features of a natural environment, that is it will represent key aspects of a real aeroplane being really flown by a real pilot. The relations between the appearance of the simulated elements and the pilot's operations on them, purely internal as they may be, have to match types of relations that would hold in a real environment. A picturing relation of the iconic kind, as discussed earlier, enriched with some indexical elements and interactive opportunities, will be the aim.

To the extent that such types of relations hold in simulated environments, and to the extent that the information involved in the context of usage is ambient rather than delimited, as it would be in a picture, an informational environment will not be altered but first *created*. Within its context of use, it will comprise a defined set of mapping regularities and criteria of detectability, relevance and reliability of its own. Perception of and within that environment parallels the perception of, and interaction with, a natural environment in relevant respects. The condition of convergence is replaced with a condition of *isomorphism*. Isomorphism, in this context, assumes a technical meaning that partly echoes the notion of isomorphisms in scientific modelling introduced by da Costa and French (2003), where it refers to model and target sharing a set of structural properties of material or formal kind, which are spelled out in terms of "families of relationships" between them:

(DS-2) Isomorphism is achieved if, for each domain D_a of a type of signal $\mathbf{r}_a$ that is emitted by an artefact and observable for *R*, there are well-defined structural mapping relations established, first, between r_a signals and conditions at a s_a that are internal to the artificial structure and, second, between these relations within D_a and those within a general type of natural domain D_n, where the regularities that govern mapping relations *within* D_a and D_n respectively may be independent of each other.

Similar to what holds for convergence, an isomorphism may well be partial, but the analogies that have been selected are supposed to match those properties of the target which are essential to how the receiver tracks, or would track, conditions in the target. For an interactive simulation, that match will have to be created on the levels of both perception and interaction. By implication, and in departure from the condition of convergence, which is more liberal on this issue, key aspects of the phenomenal character of what obtains in the target will have to be preserved, and hence any intrinsic information, too.

If, in a simulated environment, some object affords grasping or turning, and if it is not a physically implemented button, lever or other tangible interface, it will have to provide the right transformations of that affordance to differently sized and differently disposed users to map the analogous body-scaled affordances in real environments. Providing that affordance will mean that the simulated object will have to be graspable like a real object at least to some degree, raising the issue of how to simulate tactile and other sensorimotoric feedback. In fact, one of the main problems of acting within a virtual environment is that one's visually apparent locomotion within that environment has no counterpart in proprioception, as one typically remains stationary. This lack of isomorphism between D_a and D_n might lead to irritation or even nausea – an effect that appears to actually *increase* with the degree of realistic rendering of a natural environment (for a survey of this issue, see Sharples et al. 2014).

For isomorphism to hold, the perceivable elements of a simulation have to provide a wide and consistently structured spectrum of analogies to types of world affairs. Any mapping between a signal within the simulation and a concrete world affair will, first, occur in indirect fashion, in terms of the internal mappings within the simulations being mappable onto world affairs. Second, it will occur on the level of *types* of affairs on the target's side rather than concrete individual instances, unless the purpose of the simulation is to re-create a concrete event *in silico* – which normally is not the purpose of demonstrative but of a certain subclass of investigative simulations.

Returning to the flight simulator example, I will typically not be flying through a digital re-creation of *that* thunderstorm over the Drava valley last Wednesday afternoon but through a general-level analogue of a thunderstorm such as that one and numerous others. The analogous mapping is warranted by the overall appearance of the flying-through-a-thunderstorm simulation, as matched against conditions pertaining to any typical thunderstorm, rather than by any direct mappings between elements of the simulation and some event in that concrete thunderstorm. Any such concrete mapping (that gust of wind coming in from 330° Northwest by north at 42 knots at 14:17:53 and lasting for 7 seconds) would be coincidental, as it would not preserve the mappings between transformations of source conditions and signal required for convergence. Antecedent conditions may vary indefinitely. Nor can the regularities governing the respective transformations be expected to correspond. The rendering of a thunderstorm in a flight simulator does not necessarily comprise a model of the physics of thunderstorm but only will have to partially re-create its perceivable effects. What counts for isomorphism is that the perceivable properties of the simulations reliably map onto kinds of real-world conditions as the receiver usually experiences them, and that they do so in a well-integrated fashion.

The isomorphism condition will apply not only to the paradigm of simulational environments, being its richest and most complex incarnation to date, but also to all kinds of pictorial forms that appear to iconically refer to their subject but where, unlike in a photograph or portrait painting, that subject is not a real-world object or natural scene. That subject might be of either of two kinds: first, it might be an idealised or generalised world affair exemplified by the picture – a typical alpine landscape but not the Alps, or an average male human being but not any concrete individual. Second, the subject might be a fictitious object or scene that both retains some core characteristics of real-world objects or scenes and purports to depict them naturalistically – like that band of unicorns roaming the grassy hills on some distant planet. There will be some informational mapping detectable for the receiver, but that mapping will not meet the condition of convergence, which is reserved to an artefactual structure's mapping onto concrete, clearly individuated sources. Instead, it will provide a partial analogue of what one would or could detect in concrete instances of picking up information from concrete sources, as they are, partly or wholly, exemplified by that pictorial form. Pictorial forms of either kind, by their very nature as being delimited, cannot meet the criterion of being artificially created informational environments in their own right, as it can be applied to simulational environments. Pictorial forms can only be constituents and modifiers of a pre-existing informational environment.

Given the conceptual distinction between convergence and isomorphism that I have introduced, one might believe them to be mutually exclusive relations.

An item produced by a cognitive artefact, to the extent is informational in character, would then either be convergent with or isomorphic to natural information. However, as I will try to demonstrate in the following chapter, there are artefacts and situations in which these relations overlap, which has to do with both the information and the environments involved. Moreover, given the examples discussed so far, one might believe that the conceptual distinction I introduced is coextensive with that between "reality" and "virtuality" in human-computer interaction studies, where it is used to describe the properties of artefacts that (partly) simulate properties of an environment. However, additional examples will help to demonstrate that the convergence/isomorphism distinction cuts across the reality/virtuality distinction in some important cases, and in some important respects.

My discussion of cognitive artefacts, in view of their capacity of changing informational environments, might be met with a fairly straightforward sceptical observation: I have been selecting examples of rather novel, mostly computer-based technologies, apparently implicating that these examples are paradigmatic of what technologies of this sort accomplish – and perhaps *only* technologies of this sort can accomplish. However, first, cognitive artefacts need not be computational in nature, given the definition that opened this chapter. Second, there may be much older, pre-computational, possibly even non-cognitive and, from the contemporary point of view, rather mundane technologies that have the same kinds of effects: they add new mechanisms to the pursuit of our goals (which is what technologies do by definition); they expand the scope of human agency (which is one desired effect of technologies); and they may have the effect of altering our perception of, and activities towards, the environment to the point of altering our informational environments altogether (which is what some technologies, cognitive or other, turn out to do).

Examples to support this sceptical point would be, among others, electric light and the mechanical clock.[2] Electric light had the effect of shifting human activity patterns into the nights, not only allowing for longer working hours but also permanently modifying our perception of what the night is. It is hence no less than constitutive$_W$ to these activities. The mechanical clock was the first device to portion perceived time into measurable sequences, thus working to structure and discipline human activities in a way that enabled notions of punctuality and time efficiency that were unattainable before but now became a key ingredient of the industrial revolution. Again, the technology is at least constitutive$_W$ to the activities in question (for a classic account of this role of the mechanical clock, see Mumford 1934, 13f). In a non-trivial sense, the mechanical clock was an information technology or cognitive artefact just as digital computers are. It may well count as an extension of the mind that has altered our GS-constitutive relations to our environments. Electric light, in turn, is a technology that enabled activities that were hitherto practically impossible between dusk and dawn, and that in some respects modified conditions for information uptake, while neither being an information technology nor being entitled to the status of an extension of the mind in any serious way.

Although I admit to the point that much of what I described here has a long historical pedigree, the difference of degree that I see lies in the profoundness and systematicity of the modifications of informational relations to our environments that the digital computer, in its generality of purpose and in its many incarnations, has brought about. To the extent that digital computers *are* cognitive artefacts in the meaning discussed here – they might also serve other, non-cognitive functions after all – they appear to be the most potent cognitive artefacts to date. Once these cognitive-computational artefacts became directly concerned with, and directly involved in, human interactions with human environments, the question had to arise of what they effect with respect to these relations.

Moreover, a difference in kind to the technologies of old can be identified, first, for those artefacts which are capable of partly interpreting and adaptively behaving towards the actions of their human counterparts. A second difference in kind can be claimed for technologies that create simulated environments and artefacts that blend natural and simulated elements in real environments, where the convergence/isomorphism distinction introduced earlier might turn into a rather complex affair, particularly in the light of the reality/virtuality distinction.

If these technologies succeed in either of these ways, their users' informational environments will have been modified in a significant respect: if cognition is an embodied activity that is coupled with a perceivable environment, and if that environment is enriched with embedded artefacts that adaptively and partly autonomously provide information for human activities, such an environment will supposedly become informationally more transparent and interactively more tractable. Whereas, under natural conditions, an environment cannot per se considered friendly, an intelligent environment of the said kind will present itself as a seamless, smooth and cooperative surface for its users' activities. It will affect the choice and pursuit of those activities without necessarily being perceived as having been modified in a certain way. The nature of such an environment may come to be perceived as a matter of course. However, the discussion of more detailed examples in the next chapter will suggest that this is a limiting case, and that coming to be perceived as a matter of course, if accomplished at all, is a complex and two-way affair.

Notes

1 Among Peirce's scattered and complex references to these semiotic concepts, the *loci classici* are Peirce (1868) and Peirce (1998).

2 I seem to remember that these two instructive examples were brought up by Martin Hitz in a discussion several years ago. Even if my memory should fail me at this point, my acknowledgement to him will not be misplaced.

9 The intelligence of environments

> I consider the notion of a "user interface" inappropriate. The notion of an interface suggests that we are dealing with a property of the artefact, when we should actually keep the entire interaction in view. For example, is a table's surface its user interface? Or perhaps the palms of the hands of the user touching it? How should we decide?
>
> (LFE-b-2015, translation Hajo Greif)

> As a matter of fact, technologies barely modify the world; the ways in which human beings interact with technology do. The difference may seem small but it is fundamental. The public view is too much focused on devices to be able to grasp the force of technology, when it should look at human beings and their behaviours instead.
>
> (Lobo 2016, html, translation Hajo Greif)

These two caveats concerning the appropriate perspective on artefacts have not been articulated by prominent philosophers of technology but by an ergonomist and a technology blogger respectively. This may be precisely why they appear more pointed and better grounded than what many scholars in the philosophy and social studies of technology try to express in so many words.

The relationship between human beings and technological artefacts on 'micro' (first quote) and 'macro' levels (second quote) is a key concern, first, in the largely sceptic – and often utterly romantic and conservative – German tradition of philosophical inquiries into technology, from Karl Jaspers, Oswald Spengler, Martin Heidegger, Günther Anders and Ernst Cassirer to its most recent and congenially baroque and sophisticated incarnation in Gernot Böhme (2008). Second, and from a different epistemic and ideological perspective, a classic theme in science and technology studies is the relationship of "mutual shaping" between artefacts and social practices (as paradigmatically stated by Pinch and Bijker 1984 and Edwards 1994). Without sharing either the first tradition's conservatism and scepticism about technology and the human condition or the second tradition's relativism about the human condition as such, the point concerning cognitive artefacts that I broadly share with either tradition: what is relevant is not the artefact itself or any cognitive, societal or other 'impact' it may be claimed to have but the interaction between the capacities of the artefact and human purposes and abilities on constitutive$_W$ and, sometimes, constitutive$_S$ levels.

With that caveat in mind, I will now turn to selected examples of contemporary cognitive artefacts, and try to analyse their properties in the light of the concept of informational environments introduced earlier, so as to identify the informational relations they bear to these environments, using the convergence/isomorphism distinction established towards the end of Chapter 8 where applicable. These examples will, for the most part, be introduced in illustrative manner rather than having the character of proper empirical studies. One participant observation and interviews with nine computer scientists, engineers and ergonomists are the modest empirical grounding for the discussion of some examples.[1] Interviewees were based at the Institute for Cognitive Systems (ICS), the Intelligent Autonomous Systems Group (IAS), the Augmented Reality Research Group (FAR), the Chair of Ergonomics (LFE) and the Institute of Micro Technology and Medical Device Technology (MIMED), all at Technical University of Munich (TUM). Interviews are referenced by department acronym, counter and year, for example "LFE-a-2015".

The sample presented here will not be representative of all advances that are being made in cognition-related information technologies. Nor will the examples chosen capture more than a few moments in time of the many, and often rapid, developments within a variety of fields of science and technology. The descriptions and analyses will certainly look dated within a few years' time, with a slim chance of having an afterlife as interesting contributions to the history of science and technology.

Still, the examples chosen should give an idea of recent approaches to systems that create or modify informational environments and that are capable of tracking variables in these environments, in some cases including the activities of their human users. Not every example will meet these specifications to the same extent, and in the same way. Some of the systems are not made to interact with human beings, while developing functional analogies to organic adaptations to their environment (evolutionary and cognitive robotics). Some of the systems under investigation create closed informational environments, whereas others create open informational environments (Second Life and Mobile Massive Multiplayer Online Games respectively, henceforth referenced as SL and MMMOG). One will expressly strive for human-likeness of the machine counterpart of an interaction (social robotics and embodied conversational agents), one will do so in less unequivocal fashion (SL), whereas all others will not. Some are simulational in character (SL), whereas others are only partly so, in intriguing ways (especially Augmented Reality, henceforth referenced as AR, but also MMMOG) – but the distinction between "simulated" or "virtual" and "embodied" will turn out to be less sharp than expected to begin with. Some are readily classifiable as cognitive artefacts (AR in particular, but also conversational agents), whereas others are not. Some belong to the realm of fundamental research (especially evolutionary and cognitive robotics, but partly also social robotics), some to the realm of emerging technologies (embodied conversational agents and AR) and some to the realm of established application-oriented technologies (SL and, quite recently and also very successfully, MMMOG). Some will follow some paradigm of AI (to different degrees, cognitive robotics, social robotics, and AR), whereas others will not. What emerges, I believe, is an instructive mosaic image of what informational environments are, and what the interplay between informational environments and cognitive artefacts is.

Evolutionary and cognitive robotics

The research field of evolutionary robotics addresses the question of how adaptive patterns may be established – or rather, may establish themselves – in the course of some evolution-like development.[2] In contrast to behaviour-based AI, with which it shares the notions of decentralised processing and of bottom-up modelling of basic adaptive behaviours, there is no pre-defined task whose fulfilment is broken down into smaller sub-routines which, in self-organising fashion, constitute the global behaviour. In evolutionary robotics, the self-organising characteristics are located at an even more basic level than that, namely in random variation of an artificial "genome" that produces behavioural mechanisms, and quasi-natural selection of the behavioural patterns, as the corresponding "phenotype", in the robot's environment. The robot will be moving within a simplified artificial environment, and he will be provided with only a few optical sensors that are coupled with his actuators by means of a rather modest neural network as his control system. The robot is *not* equipped with inner models, maps or images of his environment.

In a classic experiment in this field (Harvey et al. 1994), a population of robots encounters rectangles and triangles in its environment, and is supposed to evolve an ability to distinguish between them. The robots are neither provided with the task of making that distinction, nor are they equipped with sensors that could distinguish between rectangles and triangles *per se*. They can only distinguish between straight vs. oblique orientation of edges. Instead, a fitness function is defined for them that consists of rewarding the robots for turning away from the rectangles and moving towards the triangles, and punishing them for moving towards the rectangles. The algorithms responsible for the robots' behaviours vary in accordance with a random function from generation to generation. In repeated rounds of the experiment, and hence over a number of "generations", the variant algorithms are selected on the grounds of how they link the robots' sensory input to their behavioural output. Only algorithms for behaviours that, in practice, map onto the distinction between triangles and rectangles with some degree of reliability will be reproduced in the next generation. If a robot fails at tracking the right *object* at one stage of the experiment, the algorithm will not be reproduced in the next generation, where a different variant, producing a different set of behaviours, will get its chance.

However, the activation state of the sensors typical for encounters with triangles will also obtain when approaching a rectangle *from some specific angles*. Only some variants of behaviour, namely taking a partial turn and thus positioning the sensors in a different angle towards the object, will allow the robots to attain the disjunct activation states for their sensors that are correlated with the difference between triangles and rectangles. The variant algorithms resulting in this behaviour will be positively selected. In a number of runs (or "generations") of this setting, the robot's control system will have learned to *practically* distinguish triangles from rectangles with a high degree of reliability, thus establishing an adaptive fit between his behavioural mechanisms and the features of his environment that serve as his fitness function. A very basic functional analogy to the development of adaptive behaviours in natural organisms is thus established.

Still, despite the capability of practically distinguishing between triangles and rectangles, even the most successful robots will not acquire even the most basic of concepts of triangles or rectangles. Instead, the robots' behaviour is directly guided by the natural information pertaining to the shapes of triangles and rectangles, to the effect that the difference in shape between them constitutes the only relevant variable in their informational environment. The invariant tracked by the robot is the angular orientation of edges that covaries with the presence of triangles and rectangles, for which there exists one ambiguous state that has to be, and can be, disambiguated by moving in relation to those edges, so that transformations of the robot's relative position match transformations in the detected angular orientation of the edges.

If there are affordances, in Gibsonian terms, that would be provided by that one trackable invariant, these could be circumscribed as a drive-to-ability or favourable proximity to triangles and avoidance of rectangles respectively, despite the absence of concepts of triangles or rectangles that would be available to the robot. To the extent that seeking the proximity of triangles and seeking distance from rectangles is the fitness function for the robots, activities of information uptake that match this function will be activities of furnishing the respective affordances on the most basic level.

The robot's informational environment is composed of the one relevant variable that he is capable of tracking, namely the difference between triangles and rectangles. Given the experimental setting described earlier, he is bound to the domain in which the difference between angular orientation of edges, given a certain relative position of his sensors towards them, reliably and regularly covaries with the difference between triangles and rectangles. That difference being the one and only ecologically and selectively relevant condition to him, his ecological and selective environment is composed of that spatio-temporal region in which the correlation between angular orientation and the objects in question holds, which happens to coincide with the room and the time in which the experiment is conducted – unless tampering with the fitness function becomes part of the experiment.

However, the experiments in evolutionary robotics described here are not supposed to provide models of ecological perception in the first place, but to provide, in elementary embodied fashion, a model of processes of biological evolution, with the elements of variation, selection and adaptation. These processes are that model's primary target system. Notably, neither the specific variations produced by the algorithms nor the specifically relevant properties of the environment nor the behaviours of the robot itself can be fully predicted by the experimenter. Elementary cognitive abilities are part of this endeavour only to the extent that they are part of the evolutionary story that is modelled here, and to the extent that the general method allows for scaling up towards more complex, at least minimally cognitive tasks – which has been the topic of later experimental work (as, e.g. reported in Beer 2003; Beer and Williams 2015). The issue of scaling and its possible limits has been raised both with respect to the practical difficulties of scaling (e.g. by Nelson 2014) and as a more fundamental objection to all sorts of behaviour-based, bottom-up, representation-free approaches to adaptive behaviour as the foundation of cognition.

A markedly different approach to building systems that use natural information in pursuit of producing an analogue of natural adaptive behaviour is taken by what is called "cognitive robotics" by its proponents (ICS-a/b-2015).[3] Starting, as it were, at the other end of systemic complexity but omitting the evolutionary aspect that would provide a foundation for genuine functional analogies, this approach commits itself to viewing cognition from a systems perspective. This means that cognitive systems are considered as integrated wholes, so as to include in any model as many aspects of their embodiment as practically possible.

The rationale behind this perspective partly parallels the ecological approach to perception discussed in Chapter 3. It commences from the critical observation that the cognitive sciences in general, and many computational and robotic approaches in particular, are too narrow in their focus when considering only one or a few components of a cognitive system in isolation, in a detached and typically rather static experimental setting. Moreover, it is maintained that the relevance of touch and other senses and of mechanisms of bodily feedback has been underestimated in much of (very much vision-centred) cognitive science. A plausible model of a cognitive system, according to cognitive robotics, needs to account for its bodily and environmental contexts and the interplay between them, in view of the integration of various sensory modalities.

Models of this kind are realised as humanoid robots that incorporate various human-like physical traits and a broad range of sensor input, including vision, touch and key modes of proprioception such as balance and kinaesthesia. For example a visual system, human or robotic, will not be able to stabilise its view on its own when moving or being moved. Stabilisation requires physiological feedback patterns that keep the eyes or cameras fixated on the object of attention. A robot visual system will also have to replicate the functional roles of fovea and periphery of the visual field in human vision. In order to provide tactile feedback, the robot will be equipped with touch-sensitive artificial skin, and it will use accelerometers to provide kinaesthetic feedback (Mittendorfer and Cheng 2011; Wieser et al. 2011). Cognitive robots that functionally integrate these kinds of capabilities serve as a research platform to generate plausible neurological models of how human beings perform the same set of cognitive tasks.

Hence, the kind of humanoid robots involved here is humanoid not for the sake of creating human-like effects on a behavioural or interactive level (as the social robots to be discussed in the next section will typically be), but with the aim of being genuinely anthropomorphic in its functional architecture. Relevant features of the human being as an organism are replicated in reasonable functional detail. Such a robot, like evolutionary robots and all other robots built under the paradigm of behaviour-based AI, will not have to be fed with explicit models of the environment or the situations in which to act. If such explicit models were used, any attempt at sensu-motoric integration would lead to combinatorial explosion. Instead, a robot will learn to cope with its environment, partly by observational learning and partly by refining its skills in the course of executing some task, where that learning is guided by a set of basic semantic rules, allowing the robot to infer higher-order regularities in the activities it observes (Ramirez-Amaro et al. 2015). Skills of communicating or interacting with human beings as intentional beings are not part of this endeavour.

A key notion employed by, and embodied in, cognitive robotics is that of "sensory-motor maps", to be understood as the mapping of cortical structures onto certain areas of the body and their sensu-motoric properties (Cheng 2014; Mittendorfer and Cheng 2011). What makes this notion relevant is the observation that processing of sensory input and motor output is not organised along sensory modalities (visual, tactile, etc.) but as an integrated cortical representation of the information provided by various modalities with respect to the body regions involved in an activity. Motor maps also account for the phenomenon that similar brain areas will show similar activation patterns when a subject *A* observes subject *B*'s activities (Chaminade and Cheng 2009). This specific property of motor maps is cited as the reason why it was possible to transfer an entire motor map dataset developed by one robot to another not exactly identical but similarly built robot who adapted that motor map within a day. In an earlier experiment, a monkey was able to control the walking of a remote robot without walking himself (Nicolelis and Chapin 2002).

A prima facie similar notion employed by cognitive roboticists is that of a "body schema" (Cheng 2014; Wieser et al. 2011), which goes back to Head (1920) and is to be understood as dynamic representation of one's posture – the shape and relative position of one's limbs – used for the spatial organisation one's activities. Unlike motor maps, however, a body schema does not make reference to cortical mapping, and hence may be organised differently. A motor map is the set of cortical structures in which one's body parts, their sensations and their movements are organised, whereas the notion of a body schema is genuinely phenomenological, qua being a representation of a perceiving subject's posture of which he or she is supposedly aware, and which he can readily use in the organisation of action. Notably, one's body schema can be modified or "extended" by using tools. An artefact, once one is trained in its use, becomes part of his or her body schema. According to one experiment, artefacts such as additional (rather than prosthetic) artificial limbs, connected via brain-computer interfaces, can be used roughly as naturally as natural limbs. Most notably, a robotic exoskeleton was used by a paralysed person to accomplish the symbolic kick-off at the 2014 FIFA World Cup Brazil, as a demonstration of the accomplishments of the "Walk Again Project" (Lin et al. 2014).

A humanoid robot built according to the cognitive robotics approach will inhabit an informational environment that is analogously shaped to that of a human being in a number of relevant aspects. Prevailing disanalogies are to only some extent a matter of the limitations of the subset of trackable information chosen for the model. They are to a more significant extent a matter of the underlying ecological relevance of the conditions to be thus tracked. This problem haunts all approaches to embodied AI bar evolutionary robotics: unless a genuine functional equivalence with organic traits can be established for a robotic trait, the make-up of the robot's informational environment will be based on a presumed, putative relevance of the conditions to be tracked. Human sensu-motoric integration has the function it has because human beings need to track ambient information in a certain integrated way, as a matter of biological necessity. Equally, an object affords what it affords to a human being on the grounds of his or her constitution and abilities, which, again, to a large part are a matter of biological necessity. One may design analogues of these necessities in similar but tremendously more complex fashion as in evolutionary robotics, but this is not part of the approach under discussion here. Still, a cognitive robot will be in the business of tracking some of the same informational domains that human beings are in the business of tracking, even in the absence of a genuine biological, somatic, evolutionary foundation – which will always be the ultimate sceptical resource for an AI critic of the Searlean or Dreyfusean slant.

Experiments in evolutionary and cognitive robotics have been feeding into an image of the human mind according to which perception and action are coupled in direct and integrated fashion, and which has gained much prominence, within and outside computer science and engineering, over the representationalist and logicist image that travelled with classical AI and its artefacts. By different but nonetheless behaviour-based means, other approaches in Nouvelle AI, most prominently 'sociable' humanoid robotics, can be found to feed into the same image. Unlike evolving and cognitive robots, sociable robots and, as their closest kin, embodied conversational agents are designed to interact with humans in shared environments, and not only shape their own but also have a chance to alter human informational environments.

Embodied conversational agents and social robotics

Embodied conversational agents are human-like virtual agents, normally displayed on a screen, with whom some form of natural spoken conversation is possible, including the display of a number of natural behavioural cues. These include adequate intonation and prosody in spoken expressions as well as gestures and facial expressions from the agent's side. However, the mode of interaction remains mostly asymmetrical to date, as input to the system in most current agent technologies can only be typed, so that the agent is not in command of vision or other sensory modalities by which to register behavioural cues of its human counterparts. In response to the typed input, the conversational agent articulates complex, syntactically and semantically well-formed sentences with gestures, expression and intonation to match, provided that the input is meaningful and semantically located within a specified range of pre-defined topics that are made known to the human counterpart.

This is, in some respects, the reverse to approaches in social robotics, where non-verbal interaction is designed for symmetry, but where the robot may not be able to understand or articulate sentences in natural language. The human-likeness of social robots is primarily located on the level of behaviour and interactive abilities of the robot, and only partly on human-likeness in appearance. One of the best-known examples of a humanoid robot designed for simulating sub-linguistic, emotional levels of human communication is MIT's KISMET, to be followed by partly more advanced, partly differently disposed models such as LEONARDO, MERZ and NEXI (the line of research to which I am referring to here was first documented in Breazeal 2002; 2003; Brooks 2002a; 2002b, with newer developments to be discussed later in this section). It consists of a desk-mounted head with some key human facial features that are animated by small electric motors: mouth, ears, eyes and eyebrows, which are connected to a visual and auditory system via a neural network. By those means, KISMET produces reactions to people's posture, gestures, facial expressions and prosody, but not to the content of what is said, which to analyse it has no means. Nonetheless, KISMET's reactions proved to be realistic and human-like enough to their human partners to allow for a degree of natural communicative interaction that is said to have lured not only its designers but also even avowed sceptics into ascribing beliefs, desires and emotions to the robot in the course of interaction (see Brooks 2002a, 149).

In some respects, embodied conversational agents in particular may appear as players in an advanced and, notably, non-blinded version of Turing's imitation game: similarity to human appearance and behaviour is the proximate aim and, at least to the human counterpart, more channels are available through which to receive information on the computer's skills in impersonating a human expert in some specified field of knowledge. Social robots offer a different take on imitation, in allowing for a more natural and symmetric interaction in terms of visual and auditory cues while not, or only rudimentarily, offering a level of verbal conversation.

These different levels of interaction seemingly match the different levels of embodiment of the agents: the physical embodiment of the social robot allows for a more immediate, embodied kind of communication that may spare the verbal level altogether, whereas the 'virtual', that is screen-bound, embodiment of the conversational agent is more remote in interactional terms and has to rely on verbal cues on the agent's side. Especially in the case of social robots, the embodied aspects of the interaction include for example cues of spatial distance vs. proximity. A human being will afford flinching to the robot when getting uncomfortably close to it, just as the robot is capable of affording flinching to the human. Such elements of interaction would be difficult to simulate for a screen-bound agent on a display that is, qua being a display, delimited in the sense introduced for pictures by James Jerome Gibson. This observation, however, does not speak against the possibility of including a capacity for using some visual cues in virtually embodied agents, let alone against adding advanced capacities for verbal behaviour to humanoid robots.

Despite the human-likeness and naturalness of interaction involved, the aim of the systems under investigation here is not indiscernibility of the agent or robot from human beings but, either proximately (for social robots) or ultimately (for conversational agents), a design that allows human beings to engage in an anxiety-free, in some selected respects, natural communication with a computer system – which is an issue of increasing societal relevance, given that demographic developments in Western societies are such that caring for the elderly will increasingly have to be delegated to partly human-like robots and agents, acceptance of whom by their users will be a design issue (LFE-b-2015, MIMED-a-2015).

On the accounts provided by Joseph Bates (1994) and Robbert-Jan Beun et al. (2003), embodied conversational agents in particular are one experimental approach to the recurrent problem of mismatching metaphors in human-computer interaction (HCI): one of the key variables in the choice of metaphors of interaction is the degree of human-likeness, visual, textual or other, with which the system is going to appear to the user. The metaphors intended in design and the ones experienced in use are often found to diverge, thus leading to difficult or failed interaction. In this context, an additional cue for the human-likeness of one's machine counterpart may enable a better matching, in that the expectation of being able to converse with the system as with a human being is reinforced. However, there are at least two instructive limitations to this strategy.

The first limitation concerns the choice of the aspects and the degree of human-likeness of the agent. Although it might seem that a high degree of perceived human-likeness would facilitate interaction, that degree does not necessarily correlate with the perceived degree of familiarity – a problem that has been discussed in HCI research under the imaginative heading of the "uncanny valley" (Mori 1970): although perceptions and attitudes towards an industrial robot normally remain indifferent, as it neither displays a notable degree of human-likeness nor is expected to look familiar in any way, humanoid robots and embodied conversational agents will be both more human-like and more familiar in appearance – until a certain point: modest similarity is conducive to familiarity if the appropriate human features are chosen for imitation, and if the degree of approximation is well-calibrated. Robots and agents may remain clearly distinct from human beings but still appear fairly familiar in interaction, if a limited but well-chosen set of cues used in human communication is implemented in a pointed fashion – very much in the sense of a good caricature. If, however, the degree of similarity increases further, yet without attaining absolute perfection, the robot's or agent's appearance is likely to become *less* familiar, to the point of being irritating or even eerie. In fact, Ho and MacDorman (2010) present empirical evidence for the assumption that the familiarity variable in Mori's argument is most accurately described as the inverse degree of eeriness (i.e. as reassurance) in the appearance of the agent. Thus, the selection of features to be imitated by an agent system, and the degree to which they actually resemble the human model, turn out to be crucially important variables.[4] Hence, convergence or isomorphism between the cues of human-likeness of artificial agents and the features of natural human beings are best conceived as obtaining in relation to certain key variables that shall be faithfully matched, whereas others are deliberately omitted or even distorted, arguably even if a faithful match could be accomplished for them.

The second limitation with which the strategy of human-likeness of agents is confronted lies in the observable irritations in the course of interaction that are induced by the asymmetries of the transmission channels available to either side, and by the human interrogator's very knowledge of the machine nature of his or her counterpart. Instead of interrogators trying to reveal the agents' nature as a machine, as in Turing's imitation game, human users can be observed to deliberately and playfully test the limits of the agents' conversational capabilities – for example by teasing them (for the following observations, see Krummheuer 2008; 2009): in human communication, teasing is a means of applying criticism in a playful, tongue-in-cheek fashion that tests for the counterpart's sense of humour. It is a practice that relies on a degree of mutual familiarity, and it is not normally applied on first encounter. Otherwise, teasing behaviour will appear inappropriate or even transgressive. Teasing a conversational agent on first encounter thus appears as an awkwardly hybrid kind of interaction. It implies a wilful transgression, while generally remaining within the bounds of interpersonal communication and the mutual expectations that come with it: to which extent does the counterpart cooperate in the teasing game, and at which point will irritation, embarrassment or annoyance ensue? It can be observed that, first, the agent will appear irritated quite quickly, obviously not 'getting the point' of the tease and, second, that the interrogators are likely to exploit the aforementioned communicative asymmetries of the interaction to achieve precisely this end. Neither is likely to happen in normal interpersonal communication.

In the social robotics case, one will not find instances of teasing *sensu strictu*, as this practice will require an elaborate degree of verbal behaviour from both sides. What one will find, however, is that, in the set-up of the interaction experiments, behaviours towards KISMET, the paradigm-defining social robot, were actually designed to resemble the behaviours one would normally display towards a small child, with utterances of whatever content being produced in a characteristically pointed praising, scolding, comforting or other tone that is typically reserved to parent-child communication. In effect, however, the subjects' behaviours go beyond this mode of communication, in being marked by a lack of distance and a sense of not taking the counterpart seriously that would be unusual in any kind of interpersonal interaction, whereas the basic patterns of interpersonal interaction remain intact. Hence, a partial analogy to teasing can be observed. For example in a video recording of an interaction in the robotics laboratory, one subject will be found asking KISMET in a scolding tone: "Where did you put your body?" Although obviously not expecting a verbal response, the subject will be seen eliciting an apparently embarrassed expression from the robot. Another subject can be observed turning towards the laboratory staff and jokingly grimacing towards them after having successfully elicited a similarly embarrassed reaction on scolding the robot.[5]

In either example, no degree of naturalness, familiarity and human-likeness in the agent's design will ensure that it is ultimately interacted with as a person, yet neither is it unequivocally *not* being treated as a person. The agent is treated like a human being to a very peculiarly circumscribed extent that has no genuine counterpart in interpersonal communication in everyday, real-world contexts. It will be thus treated on the grounds of a selection of specifically presented elementary human-like features. The domain of signals that typically pertain to interpersonal interaction has become ambiguous in its extension. Although the human beings involved will readily distinguish between an agent or robot and another human being, there will be some relevant cues emanating from the agents and robots that cannot be disambiguated by the human counterparts to a degree sufficient to confidently place them within either realm. The agents and robots are somewhat precariously situated in human informational environments, where the boundaries of the domains of signals of being human are being tested, and may be modified, in interaction.

With respect to the convergence/isomorphism distinction introduced in the last section of Chapter 8, both conditions apply to an – instructively – limited extent: to the extent that naturalness of interaction between human being and robot or agent is successfully achieved, and to the extent that the agent's or robot's behaviours are governed by regularities, that is programmes, that have such naturalness as their aim, and to the extent that the robot or agent's expressions provide recognisable and reliable analogues of what a human counterpart would display in the same kind of situation, the condition of isomorphism holds. Under this proviso, an agent or robot will be able to provide a general model of the elementary cues by which human beings recognise a human counterpart's surprise, sadness, embarrassment or other affective state, without implying a claim that such states are actually implemented in the robot. For example a robot model was used to identify the cues by which a person's trustworthiness is recognised (DeSteno et al. 2012; Lee et al. 2013). However, to the extent that irritation and ambiguity inadvertently arise in interaction, the isomorphism will remain incomplete.

Conversely, if convergence were to hold on the level of human-likeness, the human-like cues emitted by the agent or robot would, first, have to map onto traits with human-like functions of the agent or robot itself. Second, they would have to be detectable for the human counterpart with sufficient reliability, for being governed by the same regularities. Hence, convergence would require a functional analogue of surprise, sadness, embarrassment or other affective state to hold for the robot or agent, and hence a partly human-like cognitive architecture, which is not necessarily part of the design of these systems – but see Breazeal et al. (2009) for a study in pursuit of such a design within the social robotics paradigm. To the extent that the cues involved are tailor-made to leave a partly human-like impression, and to the extent that irritation and ambiguity may arise in interaction precisely for the partialness of that impression and the non-identity of underlying functions in the robot or agent, convergence is not achieved.

A different kind of convergence-requiring situation arises when human and robot are supposed to work in coordinated fashion on a shared, environment-bound task, where it is the information on that task that has to match in the first place, and where any verbal or behavioural cues exchanged stand in the service of accomplishing that task in the first place (Shah and Breazeal 2010). In the currently most advanced approach to this kind of informational convergence in a social robotics study, a robot was enabled to purposefully manipulate the human subject's mental states in the course of interaction (Gray and Breazeal 2014). This task was accomplished by modelling human cognitive states in real time by proxy of the (intended or unintended) effects of a given interaction, and by accounting for the human counterpart's perspective of that situation. For example the robot's task was to grasp whether his human counterpart still sees some object, or whether he may still believe that it is present, after it has been moved. The resulting model was then used for cooperative or deceptive purposes by the robot. The common reference point for both partners in that interaction is the environmentally embedded information that is or would be available to the human subject – and it will remain so even if the robot's task is to deceive him or her, and hence to make his or her informational environment more opaque.

Second Life

By its very nature, *Second Life* (SL) is a testbed for emerging modes of mediated interaction between human beings within a genuinely artificial environment. All of these modes revolve around some kind of impersonation by means of controlling avatars that inhabit this environment. Within a few years of operation since 2003, *Second Life* became the paradigm virtual world, and at any rate it is second only to *World of Warcraft* in terms of membership and server space (although its popularity has been waning after the SL hype's culmination in 2007). However, an ambiguity in its ontological status already becomes palpable in attempts at classification. While, at first sight, *Second Life* may be filed under Massively Multiplayer Online Role-Playing Games (MMORPGs), again along with *World of Warcraft*, it is also being referred to as a multi-user virtual environment (MUVE; Mennecke et al. 2008). Most likely, it is both at once. What it is to its users or "residents" does ultimately depend not as much on definitions or pre-existent rules as it does on the ways in which the residents play or act in this world, and the constraints imposed by the design of that world. SL may count as a particularly malleable informational environment of the simulated kind.

Most basically, the environment of *Second Life* is a three-dimensional virtual space rendered on a two-dimensional screen. In this space, besides moving around in all directions by using a menu, the keyboard or on-screen buttons, a number of activities is possible: In the first place, an avatar will have to be chosen that moves and is interacted with in SL. Its appearance may be changed in a multitude of ways, and a profile may be created describing the SL persona's interests but, optionally, also providing information about his or her real-life (RL) self. In terms of in-world communication, local chat and instant messaging functions are available, which are normally operated by keyboard, as well as social network functions are present, including the options of "friending", creating and joining groups – however, always under the veil of the avatar's identity. An activity particular and, in fact, constitutive, to SL consists in creating digital items, such as buildings or clothes or bodily extensions ("attachments") or even entire bodies for avatars, from "prims" or "primitives", that is from basic computer-generated shapes and textures. These structures can be placed on land purchased or rented by the user, or they can be bought and sold between users with in-world currency, which is converted to and from real-world currency. This sort of activity is constitutive to SL because *all* in-world objects and structures are user generated. The same partly holds on the behavioural level, where there is a possibility for scripting bodily movements, expressions, poses and gestures for one's own or other avatars' use that are more diversified and refined than the default settings.

Access to that virtual space is granted by registering as a "resident", which does not necessitate the acquisition of virtual land. However, not only is the entire visual appearance of SL besides being a navigable space user-crated, but all real-world economic revenue for SL is generated by those users who engage in building, scripting, and then buying and selling digital items. Within this framework, and depending on users' interests, areas are designated to certain themes, from being a virtual university to being a virtual shopping mall or nightclub, or to certain theme-oriented role-playing games. By such means, well-defined places are created within the digitally simulated space. Unless restricted by membership or by age constraints for the use of "adult" content, residents are free to move to any kind of place and to engage in any kind of activity appropriate to the respective place. Only at this stage, under such locally bounded conditions, place-specific rules can be imposed and, sometimes automatically, enforced by a place's owners. Such rules are usually explicitly communicated to the resident entering the respective place. For example 'physical' attacks, that is scripted behaviours targeted at another avatar's integrity, are forbidden in most places, but are a core practice in places dedicated to fighting games. This specificity of SL serves to distinguish it from online role-playing games that are, *ab initio* and globally, dedicated to one certain set of rule-sanctioned roles and activities.

Attempts have been made to classify SL residents by their type of engagement, but not without difficulty (Bartle 2003). However one may choose to subsume residents under certain types, one very basic distinction seems to be possible: there is an apparent difference between role-players, who predominantly move within spaces dedicated to the game of their choice and who are careful to play and remain within the character assigned to them, and a more amorphous and casual set of residents who assume a variety of identities under a variety of conditions, and who do not always distinguish between SL and RL to the same degree. In the context of a role-play, any remarks that are "out of character" must be indicated as such, and sparingly used (see Sixma 2009 for an instructive example). In the wider context of SL, roles are either not defined or vary from place to place, allowing for a significant amount of communicative confusion – which may actually be part of the fun to some players.

However, even under the role-playing paradigm, a number of common interactive oddities requires residents to step out of their role, for example when having to ask for help or when offering advice to fellow residents who do not know how to use a certain script. Thus, the resource of interaction becomes the topic of interaction, and residents quite naturally engage in conversation about the interactive possibilities and limitations of SL. Moreover, residents are intermittently confronted with technical warning messages from the server and with service disruptions, or users' computers or SL applications may simply crash in the middle of a conversation, making the avatars disappear or 'freeze'. These common mishaps have an effect on interactions, and may be turned into a topic of further interaction.

Above and beyond such disruptions in the perceived stability and continuity of the SL environment, the genuine ontological status of that environment is open to interpretation. Although referring to an ontology in this context may appear as a philosophical exercise for its own sake, its implications are quite practical: just as different places in SL have different rules, SL may be a different kind of environment to different players, depending on their interactional aims and self-perception. Symptomatic of such disagreement are arguments among residents about the extent to which they are supposed to reveal details of their real lives in SL interaction. For example the appropriateness of posting a real-life photo to one's profile, or of asking other residents to exchange RL photos or to use voice chat is very much contested, as is the question whether one's SL character is supposed to resemble one's real-life self. For example is a "women only" club restricted to female avatars or to female players? In fact, gender identity remains one of the most hotly contested issues in SL. Although it is quite common for players to explore alternate gender identities in SL (Ducheneaut et al. 2009; Hussain and Griffiths 2008), the discovery of such supposedly 'faked' identities is likely to lead to conflict between residents.[6] The most notorious cases of ontological ambiguity are to be found in the possible effects and the meaning of emotional involvement with other residents. Do residents fall in love with other avatars, or with the person behind the keyboard, or are both involved in a falling-in-love role-play? The players' answers to these questions will have a bearing on how they evaluate other players' alternate identities.

To none of these problems are there rules or guidelines of global applicability, nor are there reliable and universal cues to be found within the environment. Although the textual interaction possibilities in SL are mostly identical with that of any online chat, other kinds of behavioural cues can be simulated on different levels of perception. These may or may not be internally coherent for one character, and they may or may not be congruous between different players. SL is a genuinely open environment, suitable to many varieties of interaction, in which certain artificial channels are available for the transmission of information, while most of the natural channels used and relied on in real life are missing. In a rich artificial environment that supersedes Turing's informationally one-dimensional teletype communication, interaction consists in games of impersonation to which conventions of real-world interpersonal communication are only partly applicable, as many of the cues used in the latter are either not present at all or can be manipulated at will, while some resemblance to natural interaction will be expected in some way by most residents, but to different extents and under different interpretations. What most residents expect most of the time is some degree of informational isomorphism between what they encounter in-world and what they encounter in real life, but it is often not sufficiently clear to what extent and with respect to which domains an isomorphism is supposed to hold, and whether it is expected to hold in the same way by other residents. In some cases, for example when SL identity is expected to match RL identity, informational convergence will be expected by some residents instead of partial isomorphism.

Even the relevance of the information picked up in SL is ambiguous: one's perceptions, interactions and so forth, when they fail, will fail only within the context of SL (unless, of course, relationships that started in SL are carried over into RL), but the perceived graveness of misunderstandings or failures will depend on the residents' overall interpretation of the relevance of what they do and encounter in SL, which, again, may vary widely and actually contribute to further misunderstandings and failures. At the same instance, an interplay has been claimed to occur between resident's real life and SL self-perception, which may account for some of the blurring of identities in the games of impersonation played in SL (Yee et al. 2007). It is the very virtuality of the SL environment, and of the identities and behaviours played out in that arena, that remain contested among its residents. SL is an informational environment whose domains and relevant variables are subject to continuous, open-ended negotiation.

Mixed Reality Games

A markedly hybrid gaming environment that combines real-world environments and virtual elements is to be found in the field of Mixed Reality Games or, in allusion to the previously introduced concept of MMORPGs, Mobile Massively Multiplayer Online Games (MMMOGs).[7] After a number of earlier attempts that remained confined to prototype or niche product status, this type of mobile game has finally reached a broad audience and significant media attention in 2016 with *Pokémon Go*. These games are designed to transform everyday environments into playing fields that integrate physical modes of interaction and virtual elements in a manner clearly distinct from console or conventional online or mobile games. Using GPS-equipped mobile devices, mostly smartphones, the cues for interaction between players, other persons and features of their common surroundings are purposefully modified in Mixed Reality Games – and hence, if only within the context of the game, the players' informational environments. I will now give some detailed consideration to one such game, named *GeoBashing* in playful allusion to *Geocaching*, a similarly location-based game that, however, lacks some of the mobile and all of the virtual aspects of what is proposed by the developers of the game under discussion here (a first outline of the *GeoBashing* concept is to be found in Dieber et al. 2010). This game, at the time of this writing, has remained in prototype status but it already incorporates key features of what would appear in the immensely popular *Pokémon Go* five years later.

GeoBashing offers two distinct gaming modes, which can be combined: "challenges" and "fights" (with further modes considered for addition at a later stage), which are designed to create a "virtual overlay to the real world". That virtual overlay, however, is not something that is pictorially displayed on the handset's screen, like it would be in Augmented Reality applications proper (see the next section on AR). Notably, the virtual overlay in this game does not include avatars but only pseudonyms by which players are identified. Players are engaged in face-to-face interaction but, unless being already personally acquainted, they might recognise each other as players only in the course of the game. Obviously, the game is supposed to be played in public everyday contexts, on the way to or from school, on an afternoon stroll and so forth, and typically not in recession from interaction in public environments. There are no messaging or chat functions incorporated in the game. Besides real-time information on the presence of other players in one's vicinity and the status of the game, the gaming interface is used only for allocating roles to locations and spatial relations, in the form of descriptions of tasks and geographical coordinates (direction and distance) related to these tasks, plus feedback on the accomplishment of the tasks, gaming points collected and rankings between players.

In the first of the two paradigmatic gaming modes, one player could define a sequence of GPS points which the other players are then challenged to track physically, by locating them with their devices while walking or running the virtual course. Gaming points ("experience points") are awarded to those players who complete the sequence first or fastest. Separately or in conjunction with such challenges, a player could pick a virtual fight with another player in his vicinity who, after being located and formally solicited, is alerted to the imminent attack and has a "fight or flight" choice at hand. Although the flight option has to be physically realised, as the targeted individual will have to run from his pursuer to bring himself out of the (rather narrow) fighting range, the fight itself is not supposed to include any physical elements apart from the players' spatial proximity. Instead, it is played out on their mobile handsets as a random function based trial of strength, with no tactics available to the players to overpower their adversary.

On occasion of a set of field trials of a *GeoBashing* prototype with high-school students and first-term IT students at the University of Klagenfurt, the author of these lines and his colleagues conducted a short series of participant observations.[8] What could be observed on that occasion was that what had been intended as a virtual overlay to the real world interacted with aspects of the real world in several respects, not all of which were anticipated: First of all, and least surprisingly, everyday surroundings, like a city park or a university campus, are transformed into a playing field, by virtue of being thus perceived and treated by the players. Being tagged with GPS coordinates, otherwise wholly insignificant and featureless locations are turned into landmarks, which to detect and physically occupy is essential to the purpose of the game. The patch of ice behind the building on the wintry campus might turn out to be the target, while the building itself might be a mere obstacle. (Actually, buildings cannot possibly be more than obstacles in this game, as GPS becomes unreliable or dysfunctional indoors, making the gameplay a dedicated outdoor activity.)[9]

In terms of perceptual coordination, the players are presented with the predicament of mapping the tasks and coordinates displayed on their handsets' screens onto what they encounter in their surroundings, hence having to keep track of both at once. In practice, this predicament is solved in strikingly variant ways. On being presented with a challenge as an individual, some players would start running toward the first GPS point indicated on their screen without hesitation, with little regard to whom or what they would encounter, and all on their own, paying no attention to other players nor to much else of what goes on around them. Others would start crowding together, discussing the challenge and trying to solve it cooperatively. A third subgroup of players (who, in the present sample, is coextensive with those participants identifying themselves as players of online role-playing games) would not pay much attention to the initial challenge, opting to spoil for a fight instead. Hence, different players would be found to treat the gaming environment differently – an observation that might change once an etiquette of gameplay is established for that game. Alternatively, and in analogy to what can be observed in SL, an emerging gaming community might develop into a variety of subcultures, with expressly different views and goals.

In terms of interaction with the aspects of the environment that are not included in the game, it has been observed that a specific kind of oblivious immersion is typical for the use of console games and online computer games played at home. However, the behavioural implications of this state of mind are at most marginally visible in public for console games. Mobile games that do not appear to involve the environment may be played with a more visible form of oblivious immersion, or alternatively in a more casual, distracted fashion. In either case, the variable of place exerts its influence in psychological and emotional ways in the first place, rather than being directly included as an element of the game (see Hjorth 2011; Hjorth and Richardson 2011).[10]

In our empirical case, it was observed that immersion in the game results in largely ignoring those aspects of the environment which have not directly been assigned with a role in the game, or partially integrating them in improvised fashion. This somewhat suspended state of relating to the non-game environment is likely to result in strange encounters of a specific sort. For example a garbage truck or delivery vehicle entering the space that was used as the playing field, and hence as some form of territory, were mostly treated as just another obstacle. The slightly bemused looks of the non-participant delivery and garbage collection workers on encountering a crowd of young people transfixed to their handsets and seemingly erratically moving about the place were largely ignored, in some respects as if the workers' and the vehicles' presence were not much else than static to the information that the players are immersed in tracking and acting upon.

A mobile online game of the sort described here will perhaps be the most clear-cut case, or even a paradigm, of a transformed informational environment. In physical terms, and viewed as an external environment, the university campus remains the university campus, whereas many informationally relevant variables are artificially added to this environment and are acted upon as if they were natural affordances. The purpose of the game lies in tracking the domains of the signals of the relevant variables, which are the geographical coordinates that are part of the challenges to be completed and the presence and behaviours of other players.

Although the relevance of the variables in question is restricted to the game – a kind of restriction that will apply to any clearly circumscribed practice – they are constitutive of the players' behaviours within that context. The information and interaction opportunities provided by the game do not concern the external environment in which players are moving but are informational with respect to the contents of the game, which may but need not be partly sensitive to conditions at the concrete location of the player. One is supposed to perceive his or her surroundings differently in a gaming situation, in some cases turning everyday objects into elements of the game, thus temporarily redefining their meaning and function. The virtual overlay to the real world introduced by the game, by creating domains of artificial information that supplement and partly displace those of natural information, structures the players' behaviours in a straightforwardly embodied sense.

The most successful commercial application sharing the features of MMMOGs described and discussed for the *GeoBashing* prototype, and adding more features to the game, is *Pokémon Go* (Economist 2016; Kohler 2016). The first significant difference to the former game is the inclusion of AR elements (discussed further in the next section) in the gaming display instead of purely GPS-based challenges. One AR-involving gaming mode consists of finding and capturing variously shaped Pokémons that are trackable in the environment on the basis of GPS data, and that are visible on the smartphone display as embedded in the player's surroundings by virtue of a camera-based see-through function: looking at some spot through the display, the player will see the Pokémon as a moving virtual item in his or her surroundings, and he or she has to capture it by hitting it with a virtual ball. A second gaming mode consists in finding so-called "Pokéstops" that offer gaming resources, again on the basis of GPS data. These stops are landmarks or more innocuous objects with an AR overlay that have to be physically approached in order to collect the resources. A third AR- and GPS-based gaming mode consists in finding "gyms" where to "fight" other players by proxy of the Pokémons captured to date. Here, the trial of strength is based on the number, kind and rarity-related power of Pokémons collected. Intriguingly, the kind of Pokémons one has available for capturing covaries with variables in the external environment: for example water Pokémons are found near water, or fire Pokémons in hot places. Gaming strategies might involve seeking out particular, and particularly rare, Pokémons in the appropriate places. In most respects, gameplay and player's behaviours are supposed to be analogous to what was observed with respect to *GeoBashing*, from the modes – and sometimes physically dangerous excesses – of immersion to the options of teaming up vs. competing in the accomplishment of gaming tasks, with one issue gaining more prominence: it has been reported that Pokéstops can be found having been set up in places, such as cemeteries or Holocaust memorials, in which mobile gaming will appear inappropriate (Akhtar 2016; Peterson 2016), prompting the provider to remove the gaming items and make the respective areas unavailable to gameplay, and prompting calls for codes of ethics for AR games (Cross 2016).

In the case of Mixed Reality Games, the issue of convergence and isomorphism of the information provided is fairly complex: informational relations pertaining to the virtual realm, that is the game, interact with natural information for perception rather than being self-contained in a virtual environment or assigned to supposedly well-defined sub-domains. The environmental variables that are supposed to be tracked by the players are elements of the gaming environment, while several non-game environmental variables remain relevant. In a virtual reality game, in a simulator environment or in SL, that environment will be closed, so that the domain of the information involved would be (largely) coextensive with the virtual world. Not even isomorphism with natural informational relations would be required in principle, although any decrease in isomorphism would make that environment look and feel increasingly bizarre. In contrast to virtual reality or simulator environments, a Mixed Reality Game environment is open by definition. The virtual elements of the game are supposed to be informationally isomorphic in terms of the gaming tasks, which have clearly identifiable type counterparts in the real world in general while being realised in a concrete environment that cannot possibly be controlled by the gaming operations. The game itself is not supposed to furnish information on real-world variables beyond identification of players and locations and the mapping of space-related tasks onto the environment as it is encountered by the players. It is the players themselves who have to accomplish ad hoc mappings between the gaming tasks at hand and the perceivable structure of the environment and the interactive requirements it imposes upon them, and who may succeed or fail at accomplishing these mappings.

Augmented Reality

In the reference that has remained canonical for decades, Ronald Azuma (1997, 356) defines Augmented Reality (AR) as follows:

AR [is] any system that has the following three characteristics:

(1) Combines real and virtual
(2) Is interactive in real time
(3) Is registered in three dimensions.

In elaboration of this first definition, Azuma et al. (2001, 34) states that an "AR system supplements the real world with (computer-generated) objects that appear to coexist in the same space as the real world [so as to combine] real and virtual objects in a real environment [. . .] interactively and in real time". By these means, orientation or interaction within the environment shall be facilitated. Hence, its primary purpose lies in the support of human cognitive tasks ("Intelligence Amplification", FAR-a-2015), and to supply information in such a way as to make it clearly recognisable without becoming obtrusive or confusing (LFE-b-2015). Established AR services include navigation, virtual guided tours and mobile gaming (see the preceding MMMOG example and the earlier navigation example on p. 139–40 in Chapter 8). Visual support of endoscopic surgery and mechanical engineering routines are cited as other promising fields for AR applications, and automotive services, such as driver assistance, are highlighted as the potential main field of application (FAR-a-2015, IAS-a-2015, LFE-b-2015).

In line with Azuma's approach, AR researchers frequently refer to Paul Milgram's continuum between real and virtual environments, in which AR is located closer to the "real" than the "virtual" end of the continuum, within the broad realm of "mixed reality" that stretches between these two poles (Milgram and Kishino 1994). AR appears to be the conceptual counterpart to Virtual Reality in many ways: instead of cognitive and, possibly, behavioural immersion in a simulated environment, information that is not or not directly accessible to perception is added to the perceiving subject's environment (LFE-a-2015), predominantly on the level of visual perception. Such additional information may be displayed in the form of text, audio, images, graphics or video on mobile phones or other devices. In its most advanced versions, the information and interactive opportunities provided by AR applications are not confined to delimited displays but projected onto objects and surfaces, so as to become integrated with their user's perception of his or her surroundings (Feiner 2002) – which would bring all three of the aforementioned characteristics of AR to full realisation.

With respect to the intended functions of AR, one should clearly not presume a contrast between reality and virtuality in which what is virtual appears as unreal, and hence incurs connotations of being fake, made up and devoid of relevance. In fact, AR applications are paradigmatically supposed to help to provide information about the real world to their users *by means of* virtual elements, some of which may well refer to real objects in their user's environments, and none of which are supposed to tamper with correct and effective perception of the non-virtual components of the environment – although they might end up doing so under some circumstances. Only in gaming-oriented AR, as discussed earlier (MMMOG), might the first of these two conditions be waived.

The reality/virtuality distinction in AR is supposed to highlight differences in the mode of *presenting* information. If it is displayed on a screen or other delimited interface, the locality- and context-sensitive information provided by an AR application remains clearly perceivable as having been added to the environment, and it is still unequivocally localised on an interface that is itself an object in that environment, equipped with real or virtual buttons, keys or bars for interaction. In the most advanced cases of AR, however, the optic array itself becomes the interface in a certain (although probably not properly Gibsonian) sense. Virtual objects, although not being present in space, are projected onto surfaces in such a way as to make them appear as integral parts of the environment. Ideally, they should also allow for some degree of manipulation by their users, which poses additional challenges for their integration into perception (Zhou et al. 2008).

If there is a mix of realities involved in AR, the reality/virtuality distinction will capture only one dimension of that mix, as it remains focused on the mode of presentation of whatever kind of information it is concerned with rather than the nature of that information. In order to account for the relation of what is afforded by AR systems to natural information for perception, the distinction between informational convergence and isomorphism may be added as a second dimension. An AR application may provide information about concrete objects and activities in the environment that are of shared concern, and it might provide interactive opportunities towards them. Alternatively, an AR system might aim at a more general parallelism between what it signals and real or fictitious objects and activities, or types thereof, such as in AR-based gaming. I will briefly present two types of concrete AR applications, so as to further explicate these relations.

The first example is driver focus technologies, that is advanced driver assistance systems. In the wake of other driver assistance technologies, for whom a 'trickle-through' effect from the upmarket to the middle segment of the automobile market is already observable, and where a process of making them mandatory by law has commenced (LFE-b-2015, citing the example of *eCall*), driver focus systems stand a chance of becoming marketable in the near future.[11] The basic function of a driver focus system is to autonomously monitor both the current traffic situation around the car and to detect potentially hazardous situations and to observe the driver's degree and direction of attention, and to match these variables against each other and in real time. Tracking the driver's direction of attention is accomplished by camera-based detection of the viewing direction and the current geometry of the driver's face. If and when a critical situation emerges outside his or her current perceptual focus, or if there is evidence that he or she is completely distracted from traffic events in a critical moment, the Driver Focus System sends an optical signal through an LED strip that completely encircles the vehicle's interior. It might also send an additional acoustic signal. The optical signal comprises intuitively colour-coded moving light bands that converge in the direction of the relevant situation registered by the system (e.g. red for imminent danger of collision, violet for close approach to an obstacle when manoeuvring). Unlike conventional proximity sensors, a situation-specific warning is issued only if and when the system actually registers a mismatch between the situation and the driver's attention.

Driver focus technologies of the kind just described thus involve a tracking of, and reaction to, the driver's behaviours and include this information into what they display to him or her. This is information that would be fairly difficult for the driver to track himself or herself, and it is practically impossible for him or her to do so from a third-person perspective. In fact, such information is displayed to the driver especially in situations where his or her self-awareness is registered as being below par. Nonetheless, a driver focus system provides real-time information about the environment, including the driver that, although partly not present and partly even inaccessible to him or her by direct perception, has all the characteristics of natural information for perception. Convergence of this information with what the driver could hypothetically observe himself or herself (or what an attentive passenger could actually observe) is paramount to the proper functioning of a driver focus system. The domain of artificially created signals and the domains of natural information have to map onto a shared (identical or overlapping) set of concrete world affairs in a shared environment, and it will have to do so along corresponding regularities of informational mapping. After all, the purpose of the Driver Focus System is to help the driver to track relevant conditions in the environment.

A failure of informational convergence, so conceived, between what a driver could perceive and what an assistance system is capable of registering appears to be responsible for the first-ever fatal accident of a semi-autonomous vehicle on a Florida highway in May 2016. According to the accident investigation report, the car was moderately speeding at the time of the accident while "the driver was operating the car using the advanced driver assistance features Traffic-Aware Cruise Control and Autosteer lane keeping assistance" (National Transportation Safety Board 2016). Notably, the car manufacturer refers to these features as an "autopilot" function (Tesla Motors 2016). The systems involved, as several reports indicate, failed to distinguish between the side of a plain-coloured white semi-tractor trailer that was crossing the highway at right angle and the bright sky in its background (Knight 2016a; Reuters 2016; Tesla Motors 2016). It appears that the driver on the highway neither tried to intervene when the other vehicle infringed his right of way nor paid any attention to the traffic situation in the first place (some witnesses suggest that he was watching a film instead, see Reuters 2016). Issues with distinguishing bright, plain-coloured objects from the bright sky were known to the manufacturer, and detecting obstacles crossing the path of the vehicle at right angle was not part of the sensor's specifications (Knight 2016b). At the same instance, the car, being a semi-autonomous vehicle without a genuine autopilot function, included a feature to monitor the driver's hands remaining at the steering wheel. This single-cue driver monitoring feature, which was not coupled with the traffic monitoring and steering system, was supposed both to issue warnings to the driver and to slow down the vehicle if his or her hands remained off the wheel for a certain period (Tesla Motors 2016), but appears to have been ineffective for some reason. Hence, the information available to the traffic monitoring system did not match the visual information that the driver would have had available if he had kept the traffic situation in view, and no information was exchanged between traffic and driver monitoring systems that could have corrected for that situation. All of these sets of information would have had to be integrated for convergence to obtain.

The second, historically slightly older example is the mobile gesture-based interface *Sixth Sense*, developed at the MIT Media Lab (Mistry et al. 2009) where, however, it did not progress beyond an advanced prototype status. The basic concept of that interface is to provide information and afford activities relevant to the context of whatever may be the user's current purposes independently of a fixed physical interface. The system consists of a small camera and a small projector that, sufficiently miniaturised, could be worn in front of the torso like a pendant, and that are networked with a mobile device operating in the background. When looking forward, camera and projector are usually equally oriented in viewing direction. With small, maximally intuitively designed gestures, objects or people can be recognised, relevant information can be retrieved from the Internet or interaction possibilities can be invoked without further user input; drawings can be created, photos can be taken and virtual objects projected onto a suitable surface can be manipulated. For example it should be possible for the user to draw a real, physical book from the shelf and flip through it, while an application recognises the book and projects meta-information (such as ratings, reviews, additional information on the content) directly onto the cover or any open page. Moreover, a wall may become a telephone dial, a display for travel information or a game board, depending on the activity intended by the user. Beyond tangible practical applications of this kind, the design leitmotif of *Sixth Sense* is that it shall "give us seamless access and easy access to meta-information or information that may exist somewhere and that may be relevant to help us make the right decision about whatever it is that we are coming across" (Maes and Mistry 2009, TED Talk).

An interface with the capacities of *Sixth Sense* projects information and interactive elements into the environment that may, but neither need nor always do, concern concrete world affairs within that environment. Hence, there will be situations where there is no information on objects or situations within that concrete environment that would be projected into it. The information provided by *Sixth Sense* may concern any other, related or unrelated world affair, such as the status of one's flight being projected onto any given wall or the back of the front seat of one's taxi to the airport. It may also concern fictional objects if gaming is involved. Hence, the virtual nature of some object is indifferent to the convergence/isomorphism distinction.

At the same instance, and independently of the issue of the reference of the information involved, a smooth integration of what is displayed with perceptual uptake of natural information in a given environment has to be accomplished, in order for the user not to become unduly distracted or irritated by the display. An AR system of the kind and sophistication of either *Sixth Sense* or the Driver Focus System is supposed to track both environmental variables and as users' perspectives and behaviours with respect to those variables, and integrate them. Tracking of either sort has become the predominant topic in AR research, and is acknowledged as one of its main design challenges (Zhou et al. 2008). Inasmuch as people normally rely on a number of different, not exclusively visual cues when tracking an object in their environment, a richer realistic impression of virtual objects in AR will be provided by hybrid tracking across different sensory modalities (Feiner 2002, 53). In practice, however, these additional modalities are typically confined to auditory and tactile feedback, at the expense both of the other senses and of genuine provision of information by non-visual means (FAR-a-2015, LFE-a-2015).

For systems that, like *Sixth Sense* but unlike the Driver Focus System, project virtual objects into the environment as related to the user, the movements of a user's body, head and eyes, and thus his or her position and movement relative to a virtual object, will be factored into the equation by which virtual object is aligned with the physical objects surrounding it. The object's changing shape when viewed from different angles has to be accounted for as well as the effects of changing lighting conditions or partial occlusion by physical objects. Thus, besides an adequate design of digital objects, a correct and real-time calculation of how to integrate an object into the users' stream of perception is part of the practical requirements for those AR systems that use the user's surroundings as the interface. Even for the more modest varieties of AR, the users' perspective and behaviours will have to be reliably tracked, in conjunction with the context in which they obtain. What is he or she looking at? What object is he or she following? What is relevant to him or her in the present context? Tracking the domains of information relevant to the human counterpart will require convergence with natural informational domains.

To cite issues that arose in trials of another gesture-based interface, albeit one more modest than *Sixth Sense* and concerned with gestural control functions for utility vehicles, it was observed (as reported in FAR-a-2015) first, that the available vocabulary of unequivocally discernible gestures is very limited with respect to the set of commands required for effective and comprehensive control of the vehicle. Second, that natural gestures inadvertently produced by the operator which are not intended as commands are easily mistaken as commands by the system – a problem referred to as the "Midas Touch" problem: just as the mythical king's wish that *everything* he touches turns to gold coming true turned against him once he reached for food and drink, one would wish to be able discern what shall be a command and what a natural gesture when interacting with a gesture-based interface. Disambiguation, then, will require invocation commands or similar indicators of the proper context of a gesture to be available to the system, so as to provide the domain of command gestures with explicit and reliably trackable boundaries.

To sum up the importance of informational convergence in AR systems: such convergence may but need not obtain with respect to affairs in a shared environment. Instead of convergence with natural information for perception, more general isomorphisms might be the relation of choice, depending on application and usage involved. However, what has to converge in either case is the information required for successful interaction between AR system and user, and hence the cues by which objects and activities are recognised. If convergence is achieved on either or on both levels, human beings and AR systems will share a focus on certain objects of reference or towards certain conditions in their common surroundings, thereby implying some kind of *symmetrical relationship*: both are situated in a shared environment of interaction, where the success of both human beings' actions and the systems' behaviours depends on shared attention to relevant conditions in that environment, and on the stability of the conditions under which they interact. If successful, the goal states of the behaviours towards an object, condition or activity may indeed be identical, in terms of a certain task being fulfilled cooperatively.

However, this observation does not imply that the same conditions will necessarily be relevant in the same way to both sides. The particular purposes of the human and artificial participants in the situation may remain at variance, and may have to be so for the sake of accomplishing a given task. To the extent that virtual objects are integrated in real environments in AR, the function of AR is to simulate, in a different medium, some of the key properties of the environment *in relation to human perception*. In contrast to attempts at imitating human-like features for the artificial counterparts in an interaction, as in Turing's imitation game, conversational agents or social robotics, the imitation involved in AR concerns features of the environment as they present themselves to the perceiving human being. The subtle additions or changes to that environment are what is supposed to enhance human cognition and action while at the same instance human beings will need to keep track of those additions and changes in order to avoid irritation. However, irritation would arise not only from incongruously implemented, unnatural looking augmentations of reality that are cumbersome to interact with, but also from augmentations so subtle that they would not be discernible in any way from real-world objects and events anymore (FAR-a-2015). If the Gibsonian account of pictorial forms is a reliable guide, the practical requirements for augmentations that approach an ideal of perfect illusions will be improbably steep though.

Either way, the user is supposed precisely *not* to be concerned with anticipating the AR system's purposes but only with the pursuit of his or her initial goals. Thus, a significant *asymmetry* between human beings and AR systems lies in the fact that the counterpart of human interaction is a modified environment, or a modification of one's perception of his environment, not something that can be addressed as a clearly individuated being. AR technologies are meant to keep track of human relations to their environment, whereas the reverse does not hold under normal operating conditions, where the presentation of near-natural affordances takes centre stage and the aim of the technology is, after all, to become ambient.

Naturalising the artificial

Throughout all examples in the previous sections, as different as they may appear at first sight, one can identify at least one common thread: by artefactual means, informational environments are modified or, in some cases, created. At a minimum, there is an interplay between the introduction and effects of artificial structures designed to make available, augment or otherwise affect the information for an informational environment's inhabitants on the one hand, and their attempts at integrating whatever change these artefacts may effect into an established framework of reference – which might thus be altered itself. Or entire informational environments are constructed from scratch, as it were – but will have to be integrated into an established framework of reference, too, as they do not exist in complete isolation from such frameworks.

The informational environment and the artificial structures are of an obviously very simple kind to the evolving robot in the evolutionary robotics example. The entire external environment for the robot is artificially created, and hence all information that will be available to the robot – yet there is no anticipation in neither that environment's nor the robot's design of how the signals of the one relevant difference in this artificial environment will be tracked by him. If successful, the robot will, over a number of runs of the experiment, develop a GS-constitutive relation to that environment, as introduced by Peter Godfrey-Smith (1996a) and discussed in Chapter 5, under which specific behaviours of tracking the domain of signals of the difference between triangles and rectangles emerge. His informational environment is made up of the signals and the differences between signals that he has to keep track of in order to prevail.

Cognitive robots, as discussed in the same section, are not in a situation of having to evolve their own informational environments, as their functional structure is intended to resemble human cognitive abilities in some important respects. Nor is those robots' external environment artificially created, as they are supposed to act within unrestricted environments. Still, an informational environment will have been created for them, precisely by virtue of simulating some of the patterns of organism-environment relations that are relevant to the human cognitive functions.

Second Life is the other example in the aforementioned set in which the entire environment of interaction is, in the fullest and most direct sense, constructed. Apart from the basic virtual world infrastructure and the general framework of modes of interaction, residents are involved in creating, along with the shape they decide to assume and the items they construct, much of the informational relations that exist in that world, and they will do so with different aims and purposes in mind. Being a simulated environment, relations of isomorphism between informational relations within the simulations and natural information are relevant, but remain underdetermined in intriguing ways. Some informational relations within the simulated environment may not even have a genuine counterpart in the real world. Given that an avatar of any shape might be operated by a human being of any race, gender and persuasion but also by a scripted agent, the underdefined isomorphisms to real-world informational relations involved here make for what may count as the most sophisticated incarnation of the imitation game, as originally described by Turing.

In all remaining examples, rather than entire environments of interaction being constructed, artificial elements are added to existing external environments. Additions might be accomplished in material fashion, as in the case of social robotics, or in simulated or virtual fashion, as for embodied conversational agents, virtual overlays with gaming purposes (MMMOG) or virtual objects that can be interacted with in non-virtual environments (AR). The elements thus added are likely to affect the informational environments of those interacting with them, either in wholesale fashion as, supposedly, in AR, or within clearly circumscribed contexts of interaction, as for conversational agents and social robots, or within a hybrid context, as in the case of gaming affordances being embedded in an everyday environment. The player waiting for the bus home may, at an instance and without changing places, be surprised with a challenge or a fight. Although, being an MMMOG player, he will be disposed to be thus surprised, and although he also might be the one to pose the challenge or to solicit the fight with the guy at the bus stop on the other side of the street, a switching between specific informational environments obtains without any otherwise perceptible change in the external environment, apart from the presence and operation of mobile handset and gaming application.

Neither the distinction between the construction of entire environments and the addition of elements to existing environments nor the different extensions of what is modified by what means will be reliable predictors of the actual scope of the artefact-based effects on informational environments. A seemingly minor addition or a modification limited to a well-defined area of application might have more wide-ranging repercussions than its original context would have suggested. In all cases that involve human beings as users, transformations of informational environments will obtain in the use of, or interaction with, these artefacts that at some points and in some respects transcend the limits of specific and well-established contexts of application and interaction. Nor can these transformations be exhaustively explained by reference to the abilities of the machines.

These more profound transformations obtain if and when first, returning to the distinction made by Robert Brandon and discussed on p. 78 in Chapter 5, the construction or addition of elements to *external* environments has lasting effects on the *ecological* environment of the actors involved. Whenever, second, some such effect is neither transient nor has already been accommodated, situations of informational uncertainty ensue whose relevance is coupled with the relevance of that effect in ecological terms. There will be no clearly delimited domain of signals of the world affair in question that the actors have already learned to track and on which they could rely. Whether the informational relations thus added or altered are regular, stable and unequivocal as such is not the same issue as whether a relation of convergence or isomorphism between the domains of the new or modified signals and the domains of natural information that the receivers are accustomed to can be determined in practice. Resolving these uncertainties is an equally practical matter.

In this respect, the clearest of the cases under discussion here is to be found in MMMOGs. It bears some resemblance to the issue of the (hitherto still partial) accommodation of mobile phone use in public contexts to the realm of socially accepted behaviours. Until the day when these games have become a normal part of everyday practice, they will, with some likeliness, be sources of irritation to non-players as to whether their gaming behaviour, being unpredictable to the uninitiated, will be considered harmless or annoying, amusing or inappropriate. The everyday contexts in which the gaming situations occur will provide no cues unless, first, external observers are able to unequivocally identify the gaming situations as such and, second, a set of implicit or explicit rules has emerged that will tell players when and where joining a fight or challenge will count as appropriate. Once these two conditions are fulfilled, the possibility of everyday environments suddenly being turned into mobile online gaming environments will be taken for granted by players and at least tolerated by non-players.

Conversely, AR might confront its users with designs of virtual objects that may or may not or not yet strike the right balance between making the similarities to physical objects appear natural enough while making the differences to physical objects obvious enough to the user, so as to both facilitate interaction with the virtual objects and allow for the recognition of their identity as virtual. The introduction of virtual objects into real-world environments is likely to require some degree of habituation and accommodation from their users, so as to put the similarities and differences in context. Complete intuitiveness of interaction might be a convenient fiction anyway, where ease of learning should be the design paradigm instead from an ergonomist's perspective (LFE-a/b-2015). The lower the degree of habituation and accommodation that is required, the less profound the changes in the user's informational environment will be, and vice versa.

If we look from the other side of the interactive relation, from the system to the user, the requirements will be partly complementary. Quite obviously, coming to correctly anticipate and interpret human behaviours will be crucial to the system's success. To the extent that an AR system falls short of correctly interpreting human perceptions and actions, and to the extent that it does not correctly anticipate the effects of its own behaviours on human perception and action, it will be confined to some kind of cognitive trial-end-error game. The resulting behaviour is likely to create irritating effects on the side of the users which will have to be dealt with but from which the system, if so capable, may also learn.

In the SL case, although one may teleport and fly in this environment, and although avatars may assume the shape of animals, robots or even potted plants on wheels, one expects to interact with other natural human beings. However, much of the information that would be used in face-to-face interactions is not present in this environment, or it is transmitted through channels that would otherwise be complemented by information arriving through other channels (a gesture or wink to see, an exaggerated intonation to hear). Moreover, SL residents encounter the entire virtual environment in which they act with a sense of uncertainty towards its ontological status that can hardly be fixed in-world. The domains of informational relations in that world, both in terms of the unequivocality and stability of the correlations involved and in terms of the degree of isomorphism to natural informational domains, remain in constant jeopardy.

In all cases under discussion here, if the conditions of ambiguity of the informational relations were selectively relevant in a biological sense, such ambiguity would not persist for long. Still, actions or entire practices may fail in painful and embarrassing ways. An evolving robot will be at a selective disadvantage when being unable to resolve the ambiguity in edge detection from certain angles, so his artificial genes are unlikely to be replicated in the next generation of the simulated evolutionary process. In the SL case, residents with different perceptions of some game-relevant conditions, after a series of communicative and interactive mishaps that they encounter as "newbies", will be found either to quickly drop out of SL altogether, or to coalesce with like-minded residents, save the occasional clash with members of variant subcultures in this essentially social environment.

In turn, conversational agents and social robots are boundary cases in which the counterpart of interaction is known not to be a human being but supposed to be human-like in appearance and behaviour in important respects. It will be found being treated unambiguously neither as a person nor as a machine – an ambiguity that may turn out to be of somewhat more general relevance than in SL encounters. Even the most natural and familiar appearance would not warrant the agent's or robot's integration into common modes of social interaction, as the interaction remains asymmetrical, and as the counterpart's artificiality remains discernible. Some of the behavioural cues by which human beings identify each other as such are present but, first, one cannot be sure that they convey information on any sort of analogue of a real human being's beliefs and desires. Second, whatever human-likeness is conveyed in such interactions remains partial in a way that allows for uncomfortably surprising experiences of un-likeness. These conditions leave some room for uncertainty in the human counterpart about what kind of being he or she is interacting with. As long as there is no strategy of disambiguation that involves the robots and agents themselves, these beings will exist in a state of suspended or partial sociality. A certain mode of perceiving and treating them might emerge over time, but there is no clear path or predictable goal. The further development of these technologies and the degree of pervasiveness they assume will be factors in the emergence of a less suspended and more clearly defined state of affairs.

For all examples involving human interaction, it can be said that, where human agents encounter stubborn informational ambiguities in their environments, and where these ambiguities concern relevant environmental variables, they will, by acting under and towards these conditions so as to achieve some degree of disambiguation, effectively modify the framework of reference for that interaction. They will hence accommodate those changes where possible and adapt to them where necessary. Partly in allusion to the biological and juridical meanings of the term respectively, namely of being adopted in a new environment as an organism and of assuming citizenship of one's host country, and partly in reference to the use of that term in philosophy, I chose to call this latter kind of process one of "naturalisation". If naturalism is true, and if we are natural beings involved in the business of dealing with conditions in environments as we encounter them, naturalisation is not merely something that we are subject to. It is also something that we do, and it is, at root, the other face of one and the same process.

Naturalisation, in this context, means trying to integrate the unfamiliar beings and environments one encounters into a common framework of knowledge about, and expectations towards, how environments and other agents behave. This does not necessarily mean that those beings are treated like any conventional social actor, that is like other human beings under all circumstances, nor that the environment is treated as a natural environment. More often, it means treating them in a manner that tests the boundaries of commonly shared expectations, so as to assign to them a place within the cognitive and practical order that one has come to know as his or her familiar environment. The domains of information on which one relies are tentatively adjusted, as are the channels through which information is transmitted and the conditions under which something is taken to be information. Thus, any human-like signal emanating from a human or a machine would henceforth only convey information about the presence of a human-or-machine, either asking for additional cues by which to make that information unequivocal, or asking for the domain of that information to be modified so as to accommodate the presence of the machines.

As the examples discussed in this chapter suggest that the complexity of technological modifications of human informational environments is increasing proportionally to the capabilities of the artefacts involved, the number and complexity of ways of accommodating these modifications should be expected to increase, too. Returning to the opening quote of this book, when Herbert Simon claimed that human beings, viewed as behaving systems, are quite simple and that the apparent complexity of their behaviour is a function of the complexity of their environments (Simon 1969, 126f), he was, perhaps unwittingly, stating a counterfactual conditional: there is no proper way of viewing human beings in isolation form their environments to begin with. In being intrinsically, irreducibly and dynamically coupled with their environments, human beings also have a hand in shaping the kind and complexity of these environments – which will, in turn, play a constitutive role in making the mind a rich, complex and adaptive affair. The human condition is a moving target.

Notes

1 In terms of methodology, the interviews were of qualitative, semi-structured kind that is common to "expert interviews", as typically employed in the exploratory phases studies that seek to gather an overview of the knowledge available in some field that is then inquired mostly with respect to its social aspects (see Bogner et al. 2009; Gäser and Laudel 2009; Meuser and Nagel 2009). The relevant social aspects with respect to computer science and related fields were scientists' and engineers' perceptions of human-technology relations and, more specifically, their views of abilities and purposes of users. In line with the format of the expert interview, the interviews were not structured by a standardised list of questions but by an outline that stated the interview's informational aims in more general terms. Interviews lasted between one and one and a half hours, and were conducted face to face, with the exception of one telephone interview. The interviews were not tape-recorded but protocolled by the interviewer, who in all cases was the author of this study.

2 I am primarily referring to research at the University of Sussex by Inman Harvey et al. (1994) and Ezequiel Di Paolo (2003). My account also draws on the exposition in Boden (2006, 1325–1327). For a general introduction to the field of evolutionary robotics, see Nolfi and Floreano (2000). For more recent surveys of the field's development, see Doncieux and Mouret (2014) and Nelson (2014).

3 The discussion of the cognitive robotics approach is based on interviews with the chair of the Institute for Cognitive Systems (ICS) at Technical University of Munich, Gordon Cheng, and two researchers at the same department, Karinne Ramirez-Amaro and Stefan Ehrlich, all of whom shall be acknowledged here. The interviews were supplemented by a survey of literature and brief demonstrations.

4 This is why I believe it is wrong to dismissively characterise experiments in these fields as "caricatures" of human features and behaviours on the grounds of the simplification and exaggeration implied (this criticism is brought forward by Suchman 2008). The aim of these experiments is not a mockery of the human condition, but an experimental inquiry into the physical and behavioural features by which people recognise each other as human beings. Even granted that humanoid robots and embodied conversational agents *are* caricatures, a good caricature always is a pointed (albeit not naturalistic) representation of certain salient properties of its subject matter.

5 The two sequences mentioned are at 1:47 through 1:53 and 1:37 through 1:46 minutes respectively in one part of the online video documentation of the *Sociable Machines* project, titled "Recognition of Affective Intent", http://www.ai.mit.edu/projects/sociable/movies/affective-intent-narrative.mov.

6 See, for example the blog debates in https://jira.secondlife.com/browse/SVC-4180? and http://alphavilleherald.com/2009/05/is-your-second-life-woman-a-real-life-man.html. As Derek Stanovsky (2004) argues, a metaphysical point can be made of personal identities in virtual reality being 'fake' by necessity – but only a few SL residents are likely to self-consciously play upon that point.

7 For that terminology, see, for example Bell et al. (2006), Dieber et al. (2010), Korhonen et al. (2008) and Rashid et al. (2006).

8 The field trials or "demonstrations" took place in late 2010 and early 2011. The groups were each sized about 12–15 people. Video recordings were taken of the trials, in which each group was provided with *GeoBashing*-equipped mobile handsets and set out to play simultaneously, and informal group interviews were conducted after the trials. In terms of social research methodology, the empirical setting was rather casual, and so are the empirical observations in this section. Acknowledgements go out to my co-researchers Oana Mitrea and Matthias Wieser – and, of course, to the *GeoBashing* developers, especially Bernhard Dieber, Thomas Grassauer and Jakob Mayring, for their interest in our interest in the game.

9 Under a much more practical aspect, MMMOGs could become pioneer applications for ad hoc networks, for reasons that are only indirectly related to the gaming purposes: as mobile games played in real time require the players' devices to be in permanent online status and, ideally, to have push channels available, the effects on battery life would be considerable. Ad hoc networking, in which these tasks could be distributed among a number of players' devices, might offer a solution to the energy consumption problem while at the same instance providing a natural testbed for exploring the requirements for mobile ad hoc networks.

10 For inquiries into the phenomenon of immersion in, or cognitive absorption by, computer games, see Brown and Cairns (2004), Chesher (2004) and Jennett et al. (2008). Agarawal and Karahanna (2000) deliver an influential study on immersive experiences with respect to information technologies in general, while Juul (2010) provides insights into the immersive nature of "casual games". With respect to mobile gaming, studies on their effect on human perception and action include de Souza e Silva (2009), de Souza e Silva and Sutko (2009), Montola et al. (2009), Parikka and Souminen (2006), and Stenros et al. (2012).

11 The concrete system I am referring to here, the *Driver Focus Vehicle*, has been revealed to the public in February 2013, see news reports Cole (2013) and Sedgwick (2013).

10 Afterthoughts on conceptual analysis and human nature

> If anyone starts talking to you about human nature/Ignore them, it means they're as dumb as their parents were/Our nature is to change our nature
>
> (King Face 1987, "Anyone")

One of the notable virtues of the analytical tradition in philosophy lies in its specific combination of intellectual modesty and conceptual rigour. Grand metaphysical systems, as they were typical of post-Kantian idealist philosophies of several stripes, and as they dominated much of modern European philosophy, now came to be viewed with due suspicion, for their very combination of intellectual haughtiness and conceptual obscurity. Instead, it was hoped, logical analysis might help not to solve but to actually *dis*solve some of the seemingly great philosophical questions, in proving their meaninglessness, as Rudolf Carnap did in his aptly titled "The Elimination of Metaphysics through Logical Analysis of Language" (1959) – which was originally published in German with a title that, more subtly, suggests metaphysics to be "overcome" rather than "eliminated" (Carnap 1931). Alternatively, as Ludwig Wittgenstein preferred to do in his *Tractatus Logico-Philosophicus* (1933), one might use logical means to *show* the meaninglessness of the seeming great questions in philosophy.

Taking the liberty of expressing my appraisal of this endeavour and its virtues by playing on the famous phrasing in the last sentence of Charles Darwin's *The Origin of Species* (1859, 490), there is grandeur in this view of philosophy – if not in every detail of content then in spirit. I will build my concluding remarks both on my appraisal and on some of the difficulties of the analytic view of philosophy, carving out a domain within my arguments in which conceptual analysis may be appropriate – while ultimately adopting a naturalist perspective in my effort to overcome metaphysics. For the most part adopting a broader, bolder, more free-form and, at times, more polemic stance than in the preceding nine chapters, I will tentatively probe for some of the possible wider philosophical implications of my project, and develop an idea of what sort of project it actually is in the wider context of the history and methods of philosophy.

A domain for conceptual analysis

Commencing from a revision of Immanuel Kant's account of the analytic/synthetic distinction (1781, "Einleitung"), the conceptual analyst's hope was that it would be possible to ground epistemology by sorting out those statements which are analytically true, that is true by virtue of the meanings of its constituent concepts and thus by logical necessity, and justified a priori, that is justified independently of experience, from a variety of other types of statements: those which are synthetic, that is true only by virtue of how the world stands and thus logically contingently, and those which are justified only by experience, and hence a posteriori – and those which Kant considered the genuine domain of metaphysics, namely the class of synthetic a priori statements. Using Kant's own examples, the statement "All bodies are extended" is analytic because the predicate of extendedness is part the definition of, and thus logically tied to, the term "body". It is impossible to think in a logically consistent fashion of something that is a body and not extended or, to use a contemporary way of expressing this condition, it is true that bodies are extended in all possible worlds. Typically, a statement that could be tested for analyticity will have the form of, or be translatable without loss of meaning into, a universally quantified subject-predicate sentence of the form "All *s* are *F*", so as to see how the predicate term *F* relates to the subject term *s*.

Besides being analytic, the statement "All bodies are extended" is also a priori because one – hypothetically – does not have to consult experience to obtain knowledge about that conceptual affair. "All bodies" is not equal to "All known bodies" but "All possible bodies". In contrast to this first example, the statement "All bodies are heavy" is synthetic because the predicate of heaviness is *not* part the definition of, and thus is not logically tied to, the term "body" (although, again, "All bodies have mass" would be). "Bodies exist", in turn, is a synthetic a posteriori statement, as the concept of a body does not logically contain or entail its existence, and as we have to consult experience to obtain knowledge about whether that existential quantification is true – which it is. "All bodies are heavy" is equally a posteriori but not true, as the property of having weight rather than mass only obtains in another object's gravitational field. Or consider the contrast between "All bachelors are unmarried", which is analytic a priori because being a bachelor *means* being unmarried, and "All bachelors are unhappy", which is synthetic a posteriori because it is a contingent claim that requires empirical validation – which will show that it is false. Analyticity and a-prioriness are typically fellow travellers, as are syntheticity and a-posterioriness, whereas synthetic a priori statements are logically conceivable and the notion of analytic a posteriori statements is a logical contradiction in terms.

The Kantian and the analytic path of reasoning about concepts parted on the topic of synthetic a priori statements (see Rey 2015): mathematical propositions such as "2 + 2 = 4" counted as synthetic a priori statements for Kant, as there is no conceptual containment relation of "2 + 2" in "4" but a principled knowability of its truth independent of experience. Kant equalled the possibility of knowledge of the truth of such statements with the possibility of metaphysical knowledge, in contrast to straightforward conceptual knowledge about the truth of analytic a priori statements and empirical knowledge about the truth of synthetic a posteriori statements. Carnap (1928), as before him Gottlob Frege (1884), subsumed mathematical propositions of the aforementioned kind under analytic a priori statements, by extending the notion of analyticity so as to comprise elementary logical relations such as symmetry, transitivity and negation. They sought to thus avoid some of the logical problems that arose from Kant's otherwise reasonable and relevant definition. By implication, 2 + 2 can now count as being contained in the definition of 4. The domain of analytic statements is thus enlarged, so as to include a number of non-trivial conceptual relations in logic and mathematics, whereas the entire domain of synthetic a priori statements, and with it the domain of metaphysics, is called into question.

On the grounds of that revision of Kant's distinction, the analytic philosophers hoped to be able to delimit the domain of philosophy with surgical precision, as the exercise of demarcating the extension of what can be meaningfully said by exclusively logico-mathematical means. One would admit to that domain only the first, analytic a priori kind of statements, and thus the propositions of logic and mathematics, plus some foundational concepts in the experiential sciences. It would have been a small but solidly founded domain for philosophy to work in, helping the empirical sciences to treat their empirical domains with conceptually sharpened tools, and committing metaphysics to oblivion with a few surgical cuts. Conceptual analysis, then, could have been a therapy for (or, in a more pessimistic vein, a cure against) philosophy – not necessarily in accordance with the surgical analogy that I have chosen but as something more akin to a psychotherapy against common linguistic delusions, as one interpretative paradigm of Wittgenstein's work has it (see Crary and Read 2000).

Alas, this hope did not come true, for various reasons, which to recapitulate in detail I cannot endeavour here. I will leave it at a few hints. It has been suggested, most prominently by W.V.O. Quine (1961), that the analytic/synthetic distinction is artificial to begin with, and itself a metaphysical proposition. In a slightly less fundamental version of that critique, the domain of statements that can be unequivocally demonstrated to be analytic a priori is considered too small to be entirely useful, and perhaps not much larger and not much more interesting than what Kant originally envisioned. It certainly was helpful to debunk as meaningless concepts such as "the Absolute" or "*der Weltgeist*" but having little more than logical tautologies such as "all *F*s that are *G* are *F*" as one's incontestable home ground might make the analytic endeavour look philosophically toothless. After all, debunking those metaphysical statements as meaningless did little to stop them from being uttered and consumed by many philosophers with abandon. Still, a popular, watered-down vestige of the analytic endeavour is to be found in one definition of the task of philosophy as being the clarification of concepts of various sorts: logical, scientific or even ordinary linguistic ones.

Given my appraisal of the analytic philosophers' quest for conceptual clarity and their dismissal of metaphysics, what status within the analytic/synthetic – a priori/a posteriori grid do I believe accrues to my central claims? In lieu of a narrative summary, I will list and discuss my claims as follows:

(H-1) Natural information is a strictly regular relation between world affairs in which a condition *r* occurs if and only if another condition of *s* being *F* obtains, with transformations on either side having to map in strictly regular fashion, too (the informational specification claim).

(H-2) An organism's environment is composed of the specific set of conditions that are relevant to the accomplishment and the subset of those which are constitutive to the presence of his biological functions, where these conditions may be partly altered by him (the ecological coupling claim).

(H-3) If H-2 and H-1 are true: an organism inhabits an informational environment that is specific to him, to the extent that a certain set of distal *s*-conditions in his environment is relevant to him and the related proximal *r*-conditions are detectable for him (the informational specificity claim).

(H-4) If H-3 is true: the presence and accomplishment of cognitive functions depends on the availability of natural information in the environment, where the availability or treatment of that information may partly depend on the presence and operation of artefacts (the cognitive coupling claim).

(H-5) If an organism's environment is specific and dynamic in character (H-2), and if conditions in his informational environment are strictly related to conditions in his ecological environment (H-1 and H-3), and if the presence and accomplishment of some cognitive functions depends on the presence and operation of cognitive artefacts (H-4), and if the presence and use of those artefacts, being constituents of the environment, feeds back into the relevant conditions therein (H-2), then there is no immutable, artefact-independent human nature (the anthropological claim).

Are any of these statements analytic? Such will be untenable to claim for most of them, especially H-5. However, H-1 is a statement that can be readily considered for analyticity. To be sure, it would be immediately faced with the objection that the term "information" is used in a different way in the majority of contexts, and that there are reasons why the Dretskean notion of information informally recapitulated here found it hard to gain widespread acceptance. However, these are issues not of conceptual analysis but of terminological convention, which may be mended by adding a marker of difference, for example by referring to information as intended here as natural information.

As it stands, H-1 is recognisably a definition of a concept. It can be transformed into a form such as "All information is a lawfully regular relation of the type $r \leftrightarrow F(s)$" or "Information is the case if and only if a lawfully regular relation of the type $r \leftrightarrow F(s)$ obtains" without loss, and then can be tested for its semantic properties. The strictly regular co-occurrence between distal and proximal conditions is part of that definition, so that any statement derived from it by means of elementary logical operations – such as its negation or the symmetry of the relation – is contained within it. Hence, the conceptual claim is that it is true that information is strictly regular co-occurrence of world affairs in all possible worlds. Irregular or asymmetric co-occurrences shall not be admitted under any conditions. Misinformation is not information that happens to be false but will be better characterised as non-information or pseudo-information. This is what Fred Dretske claimed to be the case for information, so as to make it foundational to the possibility of knowledge, and this is what I claim to be the case, even independent of epistemological considerations.

Given these characteristics, and given the logical impossibility of analytic a posteriori statements, the claim that information has the said characteristics will be justified a priori. We will not have to investigate a manifold of real-world informational relations and then proceed to an empirical generalisation that information has the said characteristics. No epistemological claim could be built on such foundations. Information, as conceived of here, is an elementary and ubiquitous relation, and cannot be otherwise.

A defence of analyticity will already become impractical for H-2: what is contained in the concepts of organism and environment and, for that matter, in the conceptual distinction between organism and environment? "All environments do not comprise the organisms inhabiting them" would be a start, but it is precisely the organism-environment distinction that has been called into question by environmental constructivism, developmental systems theory and all serious claims for the extendedness and environmental scaffolding of organismic processes. These critiques are by no means intended to be conceptual analyses of the terms "organism" and "environment" but mobilise empirical evidence from various domains of biology. Nor is this a terminological issue that could be mended by linguistic convention. We can redefine the term information, adding markers of conceptual difference to prior definitions, and then probe for the analyticity of statements about information, so conceived. We might try something similar for environments, following the distinction in Brandon (1995) among external, ecological and selective environments. However, these specifications will rely on empirical concepts such as ecological equilibria or natural selection. We cannot say what holds for natural selection and ecological equilibria in all possible worlds, although we can say many things about what holds for them in this world, on the Earth as we know it. All resulting statements will be a posteriori, and cannot be analytic.

If "environment" was difficult, "organism" will verge on the impossible. If we confine a definition of organisms to a near-tautology such as "All organisms live", we are stuck with that near-tautology (all organisms that are not either dead or not yet born are alive indeed). Otherwise, we would have to unpack that description into statements that refer to uncontested constituents of vital processes, such as "all organisms develop" and "metabolise" and "reproduce" and "accomplish homeostasis" and "respond to stimuli". Are these unmarried-bachelor or unhappy-bachelor type statements? I strongly suspect the latter, as, first, these descriptions do not denote elementary relations but include complex concepts themselves, and as, second, we can refer to known organisms and hence empirical knowledge only when subsuming these characteristics under the concept of an organism (should "are carbon-based" be part of that catalogue?). Moreover, we face boundary cases, such as viruses, for which not all of the mentioned characteristics hold, and where empirical disagreement prevails. If one is a Neo-Darwinist, viruses will count in (genes are replicated but much else is missing). If one is a vitalist, one will postulate a vital force as a metaphysical – and hence, according to Kant, a synthetic a priori – criterion that is supposedly expressed in the aforementioned set of empirical characteristics. Everything else we could do, it seems, is synthetic and a posteriori.

By virtue of being partly derived from a partly empirical claim (namely H-2), H-3 will itself be an empirical claim about what organisms in environments do with natural information. Any attempt to analyse this statement will initiate cascades of specifying and qualifying statements, many or most of which are of the synthetic a posteriori kind. "Informational environments" should be well-defined though, by virtue of comprising the notion of information as analytically treated in H-1. By virtue of being partly derived from H-3, H-4 is to be viewed in partly analogous fashion, while adding a concept of artefacts that is not further explicated here but certainly is not exhausted by an analysis of the "All artefacts are structures created by organisms" kind. Because the purpose of the artefacts involved here is the provision and treatment (to inelegantly avoid the strongly computation-connoted term "processing") of information, only a certain subset of these structures is of interest here, and some of them may not even have been created in a narrow sense. For this reason, I have been using the term "cognitive artefacts", or would suggest to refer to "cognitive resources" if natural structures are used for the provision of information.

What comes closest to the appearance of a synthetic a priori statement in my list is H-5. It includes a claim concerning the human condition and artefacts – which has been a topic of metaphysically inclined philosophical anthropologies since the early twentieth century and thus may invite the reader to interpret my statement as a derived from, or embedded in, philosophical anthropology. Helmuth Plessner's work may be considered the pinnacle of this specifically German tradition, in justifying a set of highly elaborated claims about the human condition on partly metaphysical partly biological arguments (of a neo-vitalist stripe), but Max Scheler, Arnold Gehlen and, in a slightly different vein, Ernst Cassirer have been more popular and accessible reference points for arguments about human nature and the nature of human artefacts.

According to Scheler (1928), man is a being whose nature is not exhausted by being a more diligent, artefact-creating animal, which he refers to as *homo faber*, nor is he bound to concrete environments; instead, and in explicit rejection of what the biological sciences may tell us in these respects, man is considered the only living being capable of viewing the world and himself as objects of thought and action, thereby abstracting from concrete conditions of existence. According to Gehlen (1940), man is a being that is by his biological nature incomplete, helpless and un-specialised in comparison with other organisms, and thus requires artefacts and other "institutions" as extensions that augment his capacities and develop a degree of autonomy that, in turn, requires from him a degree of subordination under them in order to provide him with a stable mode of existence. Plessner (1928), like Gehlen but in markedly brighter colours, places human beings back in the order of nature but attests to them a mode of existence that is at variance with either plants, whom he considers "openly" organised, for their lack of central organs, and other animals, whom he considers "centred" in organisation, for possessing central organs to regulate traffic with their environments, whereas human beings are "excentrically" organised, in being able to self-reflexively refer to their position in the world, which they, by virtue of that very nature, always access in mediated fashion, through artefacts, culture and social practices. Cassirer (1944) roots these human-specific modes of accessing the world in their nature as an *animal symbolicum*, and thus an organism that is not merely capable of perceiving and signalling immediately relevant conditions in its environment but endowed with an ability to form symbolic representations and concepts, from which all kinds of other abilities flow.

None of these characterisations of the human condition is contained in, or strictly logically derivable from, a basic concept of a human being but only from the various presupposition-rich conceptions devised by the authors. At the same instance, none of these characterisations is an empirical claim that could be derived from, or proven by, experience in general and biological science in particular. All of them are unhappy-bachelor type statements but, at the same instance, they are supposed to grasp something about human nature that is prior to, and in principle independent of, any human experiences or interaction, with partial exception granted to Plessner. Above all, these characteristics are supposed to be immutable at least in terms of being possessed by all beings known as humans, and only by those beings. A living being not possessing those characteristics would then not, not yet, or not anymore be a human being.

Hence, looking at H-5, the crucial difference between the claims made by the philosophical anthropologists and my project seems to be that I consider the human condition to be dynamic and historical. A rephrasing of my statement would be "There is a human nature, but it is neither immutable nor artefact-independent." To be sure, such a claim would constitute a major departure from the tradition of philosophical anthropology, as either predicate negates at least part of what the philosophical anthropologists claim. I could then proceed to a set of metaphysical claims about how the essence of the human condition is our ability to change ourselves and our environments, and happily sing along with the "our nature is to change our nature" chorus (King Face 1987) – after explicating why this is not paradoxical as an anthropological claim in the first place.

Alternatively, and more fundamentally, the difference to the philosophical anthropologists could lie in an act of throwing out the concept of "human nature" altogether and settling for a constructivist position. The corresponding rephrasing of my statement would then be "There is no human nature, so it could not be immutable and artefact-independent either." Nor could it be mutable and artefact-dependent for that matter, as it is the subject term that is negated in this rephrasing, so any further predications would be meaningless. The mutual shaping relations between humans and artefacts would not presuppose a human nature that would first enter into that relation. Instead, it would be a product of that relation. According to this image, the relation will first create the relata. Hence, the notion of human nature would be subject to deconstruction, in that there is nothing natural about human *nature*. This reading, however, amounts to a set of metaphysical, synthetic a priori claims, too, albeit negative ones that are at risk of falling victim to the paradoxes of epistemic perspective concomitant with any radically constructivist claim.

I am hopeful that the preceding chapters have created an impression that sufficiently stands out against either of these two readings. The primary, foundational difference between H-5 and what the philosophical anthropologists claim is that I am seeking to make an empirical claim about human beings as a species and about salient properties of their environments as being closely coupled in a dynamic relationship – without denying the possible deeper implications of these claims but also without either assuming to have proven or promising to prove them. Hence, the proper rephrasing of my statement in H-5 will be "It is a contingent empirical fact about human nature that it is mutable and artefact-dependent." This is a synthetic a posteriori statement that eschews metaphysical commitments while suggesting a perspective on how human nature may be best conceived of in extra-scientific, normative terms. This view of a partly self-made and changeable human nature, I will now proceed to demonstrate, firmly stands in the tradition of philosophical naturalism.

A naturalist's view of human nature and machines

Contemporary philosophical naturalism has been concisely defined as "the view that all phenomena are subject to natural laws, and/or that the methods of the natural sciences are applicable in every area of inquiry" (Boyd et al. 1991, 778) – which, as Hilary Putnam (2004) rightly observes, is a disjunctive definition or, more precisely, two definitions of different content and scope that reflect the two main understandings of naturalism in theoretical philosophy. A paradigmatic description of these two main understandings reads as follows:

> Ontologically, naturalism implies the rejection of supernaturalism. [. . .] Positively, naturalists hold that reality, including human life and society, is exhausted by what exists in the causal order of nature. [. . .]
>
> Epistemologically, naturalism implies the rejection of all forms of a priori knowledge, including that of higher level principles of epistemic validation. Positively, naturalists claim that all knowledge derives from human interactions with the natural world.
>
> (Giere 2009, 308)

This basic distinction is not uncommon and can be found in similar form in, for example Quine (1969c, 26), who holds "that knowledge, mind, and meaning are part of the same world that they have to do with, and that they are to be studied in the same empirical spirit that animates natural science". The same distinction can also be found in David Papineau (2007), who further subdivides epistemological naturalism into, first, epistemological externalism, that is the view that all knowledge hinges on empirical factors and can thus be only reliable, never certain, and, second, the assumption that philosophy should be continuous with the natural sciences. Mario de Caro and David Macarthur (2004, 4) refer to epistemological naturalism as a "methodological theme" and to ontological naturalism as "a commitment to an exclusively scientific conception of nature."

The ontological and the epistemological understanding of naturalism, although appearing quite closely related at first sight, are not necessarily interdependent, and they may or may not co-occur in naturalistic arguments and their critiques – although ontological naturalism is explicitly or implicitly presupposed by many epistemological naturalists. The key tenet of ontological naturalism is that the causal order of nature is *closed* in the sense of ruling out the possibility of supernatural or extra-natural interventions. Two sub-clauses that follow by implicature rather than entailment are that the causal order of nature is *complete* in the sense of all physical phenomena being sufficiently explainable by physical causes, and that it is *uniform* in the sense of the same laws of nature applying in all places at all times. For most authors, closure and completeness are synonymous. It might be worthwhile, however, to consider the distinction made by Eric Marcus (2005): closure, like uniformity, is a genuinely metaphysical claim about how the world stands above and beyond physics – namely that there is nothing that exists above and beyond the physical realm, or that, if there exists some such thing, it does not matter to the physical realm. Completeness, in turn, is a claim about explanations in the first place, namely that physical explanations are sufficient; they might, however, not be necessary. Completeness suggests closure but does not require it.

Ontological naturalism is at once the most general and sweeping and the least provable of naturalistic tenets. It is a metaphysical claim that is supported – but cannot be proven – by empirical evidence and philosophical argument. In conjunction, evidence and argument may demonstrate that assumptions of supernatural or extra-natural powers are *not necessary* to explain the causal order of nature and the place of life, the human mind and other philosophically relevant phenomena therein. Accordingly, even those authors who actually provide explicit arguments for ontological naturalism confine themselves to demonstrating that the assumption of completeness is tenable and coherent, and probably compelling. They do not claim that this assumption is true by necessity, let alone that it is strictly correlated with closure of the causal order of nature (see, e.g. Davidson 1980, Chapters 11 and 13; Papineau 1993, Chapter 1; or Smart 1963). One of the main reasons behind the logical positivist's reservations against naturalism, which seem to contradict his or her commitment to a scientific world view, is just that: the inevitably metaphysical and thus, on the view of logical analysis, nonsensical nature of materialism as an ontological doctrine (Stoljar 2009).

Conversely, epistemological naturalism, in all its varieties, seeks to ground knowledge in "human interactions with the natural world" (Giere 2009, 308) but is not of one mind as to whether the natural sciences are one incarnation, the paradigm or even the exclusive domain of such knowledge. On the most restrictive view, a 'hard', physicalistic reductionism of explanations in the special sciences to physical explanations prevails. The argument for this position, also known as the "unity of science" argument (Carnap 1955; Oppenheim and Putnam 1958), rests on the hypothesis that physics is the most basic science, for being concerned with the most basic processes in the natural world – processes on which all other phenomena rest. However, this claim is not entailed by the claim of a universality of the methods of the natural sciences in Boyd et al. (1991), which is more plausibly read as a weaker claim: unless biology, psychology or sociology make claims about an ontological specificity of their subject matter, their tenets and methods are, although heterogeneous and dis-unified, fully compatible with the closure and completeness of the causal order of nature.

Unless a metaphysical commitment to ontological naturalism and or a strictly physicalistic reductionism about explanations are involved, epistemological naturalism will be a fairly broad and inclusive doctrine, amounting to the minimal epistemological requirement that all knowledge is empirical knowledge and is to be achieved by the methods of all varieties of the empirical sciences, defined in the widest sense. This "broad church" interpretation was the hallmark of early twentieth century American naturalism, notably John Dewey (1929) and Roy Wood Sellars (1922; see the historical account in Keil 2008). However, apart from denying the possibility of a priori knowledge, such a minimalist position will be systematically, and possibly purposefully, vague – which does not mean that its purpose is vague.

For Dewey in particular, philosophical naturalism has a therapeutic function, albeit in a different, arguably more positive sense than Wittgenstein's: we will be better prepared to cope with the upsetting of traditional concepts, values and modes of thinking by modernity if we employ an "empirical method which remains true to nature", so as to inspire "the mind with courage and vitality to create new ideas in the face of the perplexities of a new world", Dewey argues in a paradigmatic statement of his naturalism, "The Influence of Darwinism on Philosophy" (1910, iif).

This philosophical-therapeutical argument cuts two ways, which are distinct but related: one of general philosophical outlook and one of suitable philosophical methods. Dewey argues that Darwin's theory of evolution changed philosophy, first, by opening philosophical reasoning about life, politics and morality to the idea of change – an idea of change, however, that is clearly distinct from random variance. Second, Darwin's theory "emancipated, once and for all, genetic and experimental ideas as an organon of asking questions and looking for explanations" (Dewey 1910, 9). Hence, on the first level, Dewey's naturalism offers an alternative route to the overcoming of metaphysics to what the conceptual analysts proposed, in suggesting that the quest for ultimate, immutable truths, whether metaphysical or conceptual, should be abandoned in favour of an open-minded stance that allows for constant revision. On the second level, Dewey's naturalism amounts to the endeavour of modelling the resources and conduct of philosophical reasoning primarily on the experimental, evidential and critical methods, and to some extent also on the concrete contents, of the empirical sciences. On this level, a naturalistic view will depend on the concrete contents of scientific inquiries, in Dewey's case on inquiries into evolution and cognition. Still, a naturalist view thus anchored in science does not need to amount to a scientistic philosophy – at least not to a larger extent than any other philosophical doctrine in history that relied on the state of scientific inquiry of its age.

Either way, scientific and philosophical concepts are likely to turn out to be subject to change, so that the very notion of the existence of a priori justified statements of any sort, synthetic or analytic, is called into question – with the exception perhaps of logical constants and the tautological statements derivable from them. Do we need an a priori concept of space to reasonably think and act in this world in the first place, as Kant (1781, "Transzendentale Ästhetik", § 1–3) famously insisted and James Jerome Gibson (1979, 32) denied? As Gibson's work was rooted in the tradition of the American naturalists, we should naturally expect him to cancel space from the list of a priori concepts, along with many others. But then, do we need an a priori concept of natural information as a specification relation in order to explain perception? Hence, should we expel the concept of natural information from the realm of a priori statements, too?

The relations singled out by the concept of natural information may be so elementary that tracking them by some means is indeed a precondition for any perception, activity and knowledge. Hence, *pace* Kant, the uptake of natural information may be a factual, practical precondition for forming a concept of space. On the account I presented in Part I, the possibility of tracking natural information allows for forming, but does not depend on the presence and operation of concepts in the perceiving organism. Natural information is used for purposes such as spatial orientation in organisms that cannot be expected to have concepts of any sort, perhaps not even the most elementary practical concepts or "unicepts" as proposed by Ruth Millikan (2004; 2017). These organisms do not have to be in a position to perceive anything *as* something in order to be afforded with the requisite activity by that information. In contrast, space, like time, may count as a method of conceptually ordering relations thus picked up that is available to organisms that are reliant on grasping objects and relations in their environments in complex spatio-temporal constellations, such as an object being behind, beneath or above another, or a moving object reappearing after being occluded by another. A concept of space also allows for abstracting from concrete relations of this sort by using absolute measures, thereby moving from the use of intrinsic to extrinsic information. If this naturalistic view is tenable, environment-bound information comes first, and a concept of space later, so the latter will not be accepted as being a priori.

If space is denied the status of being an a priori, experience-preceding, immutable concept, the concept of human nature is unlikely to fare better in this respect under a naturalistic perspective. It is open to revision on the levels both of preferred methods and of general outlook: inquiries into human nature on the biological level might lead to, or be guided by, a change in general outlook on the human condition, in an open-ended process of mutually informing each other. This commitment to systematic openness of the naturalist's quest has a corollary that, although not being a metaphysical statement, constitutes a constraint on the anthropological perspective to be reasonably adopted: man should be treated in philosophical endeavours in the same way as in scientific inquiries, namely as a natural being like any other, with cognitive capacities that are likely to be more sophisticated in degree, but otherwise, in their functional roles, analogous to what other animals accomplish in order to cope with *their* environments. Man should also be taken to be a natural being like any other in terms of living in, and having to cope with, changeable environments, where some of these changes are, intentionally or inadvertently, self-made.

Hence, the naturalistic anthropological claim to be defended here contrasts with the philosophical anthropologists' views as follows: it affirms the very the notion of *homo faber* that Scheler disdained, and subtracts any residual metaphysics from that notion. It affirms a deflated version of Plessner's claim that artificiality is partly constitutive to human nature, subtracting his vitalist preconceptions and proposing a deflationary, biologically more grounded reading of his notion of positionality. It affirms Cassirer's claim for the speciality of human abilities of symbolising but views them as properly evolved and hence continuous with animal signalling. It affirms Gehlen's notion of man as an incomplete being but, again, subtracts the metaphysics from it and dispenses with its negative normative connotations. After all, if artefacts are constitutive to the shape and accomplishment of human abilities, and if a coupling with the environment is constitutive to biological functions for all organisms, and the coupling with artefacts for some organisms, there will be no incompleteness once the whole picture is taken into consideration.

A naturalistic account as outlined here will have to accomplish two tasks that the philosophical anthropologists were free to waive: first, it will have to show in as much empirical detail as possible how human cognitive abilities are actually and concretely coupled with the environment, and hence how they actually *operate* in such a coupling relationship. How does the uptake of natural information work on the organic and neuronal level, and how does it map on activities? Where and how precisely do representations come into play? This is a task for an ecologically minded cognitive science, as outlined in Chemero (2009). I freely admit of having provided little in the way of such empirical detail on this level, instead confining my account to reviewing some work that has been done in that field and possibly offering some contributions to further concept- and theory-building therein. Second, a naturalistic account will have to show in as much empirical detail as possible how artefacts are dynamical constituents of environmentally coupled cognitive abilities. On this level, I am hopeful to have escaped the philosopher's chronic condition of armchair confinement with at least some success – while risking putting myself in the position of one of the proverbial blind men who try to collectively determine the nature of an elephant by each touching one part of him. What I tried in order to avoid that epistemological problem was to develop a good idea what the more capable men and women on the first level are doing, and match my account of natural information and cognitive artefacts against it.

More positively, there might be a historically grounded, indirect but still systematic justification for commencing from an account of artefacts, so as to arrive at some conclusions with respect to cognitive functions more generally. After all, the concept of machine played a crucial role in the emergence of the cognitive sciences, as two of its founding disciplines strongly relied on machine models: AI as inspired by Turing's work, especially (1950), and Cybernetics as inspired partly by Ashby's non-computational, material machine model of adaptive processes known as the "homeostat" (1960). These two different but related types of models – Turing and Ashby were concurrent and corresponding members of the "Ratio Club" and knew each other's work (Husbands and Holland 2008) – were crucial to the establishment of the notion of "mind as machine" that defined the cognitive sciences ever since (Boden 2006; Gardner 1985). One of the reasons why Turing considered the original question of whether machines can think "too meaningless to deserve discussion" (1950, 442, see Chapter 1) is that the "normal use of the words" (1950, 433) is so parochial and historically contingent that the very conjunction of "machine" and "thinking" seemed to be a *conceptual* impossibility on the normal use of the words in 1950. Part of Turing and Ashby's endeavours was to demonstrate that what first seemed a conceptual impossibility could become a matter of serious consideration and empirical investigation. If such an empirical investigation is possible, even if it has its limitations, and if such investigation informed the design of real-world computing artefacts, and even if these artefacts are not thinking machines, these artefacts will have made a contribution to redefining not only the concepts of "machines" and "thinking" but also our perceptions and practical treatment of what the nature of machines and human thinking actually *is*.

In one of the best-defined and most carefully argued sceptical accounts of AI, Harry Collins and Martin Kusch (1998) propose a conceptual distinction between human beings and machines that is very intriguing in this respect: on their account, a machine is distinguished from a human being by its inability to perform or "do" actions of one kind in particular, namely those actions which can be accomplished in manifold, materially dissimilar ways and can be identified only as belonging to the same type by reference to an understanding of their social context. To the observer, the behavioural patterns involved may look rather different on different occasions, but they are tied together by the socially sanctioned purposes that they serve. One can pay a bill in several ways, by handing over cash or by swiping a credit card through a card reader. It would be the same action in the light of the purpose of paying a bill – but only in a society where credit card payment is an established practice.

In contrast to these parochially human "polimorphic" actions (playing on the words "polymorphic" and "polis"), Collins and Kusch continue, human beings and machines seem to share an ability of performing "mimeomorphic" actions, which are defined as those actions which are tied to one and the same type of behavioural sequence, with no or rather minute degrees of variance permitted. There are only so many ways of performing a certain dance move or golf swing. Doing them different is likely to amount to doing them incorrectly. For their invariability, these are behaviours that can be imitated or simulated by a machine. Hence, machines might be able to imitate human behaviours of this kind, but they will not be able to – deliberately – take different routes to the same goal, as this would require an understanding of the purposes of seemingly variant behaviors.

In the light of this argument, a machine could succeed in Turing's imitation game in a restricted sense, by providing an imitation of human conversational behaviour, but it could not *play* that game, because playing a game would require an understanding of the practice of playing games in general, and playing *that* game in particular. However, the authors object, the machine would be unlikely to succeed in imitating conversational behaviour in the first place, as such imitation cannot be accomplished without an understanding of conversation in general and the present conversational context in particular. It would be like playing Chinese Whispers without knowing the language of the input word or phrase. Polimorphic actions, Collins and Kusch conclude, cannot be simulated by machines, because such a simulation would have to presuppose a genuine understanding of those actions on the part of the machines.

Remarkably, however, Collins and Kusch use a purely ability-based definition of human beings and machines. To them, a human being is whoever or whatever is or could be, under appropriate conditions, capable of performing both mimeomorphic and polimorphic actions, whereas a machine is whatever is only capable of simulating mimeomorphic actions. By adopting this definition, any hypothetical machine that might transgress this boundary would *cease to be a machine*, and the extension of the set of entities to be subsumed under the concept of human beings would have been enlarged. Hence, the answer to Turing's question, "Can a machine think?" would remain negative – presuming that there is a relation between the ability of *performing* actions and the presence of mental processes – but this would be so by virtue of the machines involved losing the defining characteristics of machines and assuming defining characteristics of human beings.

One could ask whether it is an analytic statement about machines that they are capable only of simulating mimeomorphic actions, and of humans that they are always also capable of performing polimorphic actions, but this is what Collins and Kusch (1998) seem to suggest, although their aim certainly is not conceptual analysis. But then, simulating mimeomorphic actions was certainly not part of the kinematics-based classical concept of machines, briefly summarised at the beginning of Chapter 1, that confined the domain of what machines can do to the movement of matter and the harnessing of energy. Even to Turing's revision of that view and his conception of what computing machinery can do, the imitation of human actions was relevant only in a circumscribed sense, namely as one of the indefinitely many possible accomplishments of his Universal Machine.

Collins and Kusch offer a conceptual escape route from one central – fallacious – implication of the basic point of the AI sceptic that equivalence between human cognitive and machine abilities is impossible in principle, a point the authors otherwise seem to share and which I may rephrase as follows: "Machines cannot think because thinking is whatever a machine cannot do." This is one possible rendering of the "moving the goalposts" machination against the notion of thinking machines – as, for example, reported by Pamela McCorduck (2004, Chapter 9) with respect to Hubert Dreyfus's critique of AI: if a machine can perform a certain cognitive task, successful accomplishment of that task, whatever it is, by definition, cannot be thinking. Dreyfus himself states his critique as the problem of the horizon of AI systems actually accomplishing the cognitive tasks they were promised to accomplish to be "receding at an accelerating rate" because these tasks turn out to be vastly more complex than AI researchers believed (1979, 5). On either version of the argument, the boundary of what constitutes the accomplishment of a cognitive task will be continuously pushed upward with the advent of more capable machines, and there is little we will have learned until we thereby either ultimately carve out a domain for a Cartesian *res extensa*, or something to the same effect, that would remain essentially inaccessible to any kind of machine simulation – or stand to find out that nothing inaccessible is left.

What I have been suggesting in this book sought to turn this line of argument on its head in order to put reasoning about human cognition on its feet: ever more capable cognitive artefacts become part of human environments in ways that highlight how we are coupled with our environments in perception, cognition and action, and they do so by modifying the ways in which we are thus coupled. These artefacts are not machines that think, nor are they supposed to be machines that simulate human thinking. This may be the wrong accomplishment to seek from them to begin with, particularly if the conception of the nature of cognition that entered into that quest is at risk of being incorrect, inadequate or inaccurate. Still, the abilities of cognitive artefacts move way beyond traditional conceptions of what machines can do in a variety of practical ways that may be instructive to inquiries into human cognition in different fashion. To end on an artefact-based metaphor other than that of machines: they may serve as calibrating instruments, in not simply measuring but helping to adjust our measurements of what sort of environment-bound activity human thinking is.

Bibliography

Adams, F. and K. Aizawa (2001). The Bounds of Cognition. *Philosophical Psychology* 14, 43–64.

Agarawal, R. and E. Karahanna (2000). Time Flies When You're Having Fun: Cognitive Absorption and Beliefs about Information Technology Usage. *MIS Quarterly* 24.4, 665–696.

Akhtar, A. (2016). *Holocaust Museum, Auschwitz Want Pokémon Go Hunts Out*. URL: http://www.usatoday.com/story/tech/news/2016/07/12/holocaust- museum-auschwitz-want-pokmon-go-hunts-stop-pokmon/86991810/ (visited on 10/01/2016).

Armstrong, D. M. (1960). *Berkeley's Theory of Vision: A Critical Examination of Bishop Berkeley's Essay towards a New Theory of Vision*. Melbourne: Melbourne University Press.

Ashby, W. R. (1960). *Design for a Brain: The Origin of Adaptive Behaviour*. 2nd ed. New York/London: John Wiley & Sons.

Auersperg, A., A. von Bayern, S. Weber, A. Szabadvari, T. Bugnyar, and A. Kacelnik (2014). Social Transmission of Tool Use and Tool Manufacture in Goffin Cockatoos (Cacatua Goffini). *Proceedings of the Royal Society, B* 281, 20140972.

Azuma, R. T. (1997). A Survey of Augmented Reality. *Teleoperators and Virtual Environments* 6.4, 355–358.

Azuma, R. T., Y. Baillot, R. Behringer, S. Feiner, S. Julier, and B. MacIntyre (2001). Recent Advances in Augmented Reality. *IEEE Computer Graphics and Applications* 21.6, 34–47.

Babbage, C. (1864). *Passages from the Life of a Philosopher*. London: Longman, Green, Longman, Roberts, & Green.

Bar-Hillel, Y. and R. Carnap (1952). *An Outline of a Theory of Semantic Information*. Tech. rep. 247. Cambridge: MIT Research Laboratory of Electronics.

Bartle, R. A. (2003). *Designing Virtual Worlds*. Indianapolis: New Riders Publishing.

Bates, J. (1994). The Role of Emotion in Believable Agents. *Communications of the ACM* 37, 122–125.

Bateson, P. (2001). Behavioral Development and Darwinian Evolution. In: *Cycles of Contingency: Developmental Systems and Evolution*. Ed. by S. Oyama, P. E. Griffiths, and R. D. Gray. Cambridge/London: MIT Press, 149–166.

Beer, R. D. (1995). A Dynamical Systems Perspective on Agent-Environment Interaction. Artificial *Intelligence* 72, 173–215.

———. (2003). The Dynamics of Active Categorical Perception in an Evolved Model Agent. *Adaptive Behavior* 11.4, 209–243.

Beer, R. D. and P. L. Williams (2015). Information Processing and Dynamics in Minimally Cognitive Agents. *Cognitive Science* 39, 1–38.

Bell, M., M. Chalmers, L. Barkhuus, M. Hall, S. Sherwood, P. Tennent, B. Brown, D. Rowland, S. Benford, M. Capra, and A. Hampshire. (2006). Interweaving Mobile Games with Everyday Life. In: *CHI'06 Conference on Human Factors in Computing Systems*. New York: ACM, 417–426.

Berkeley, G. (1709). *An Essay towards a New Theory of Vision*. Dublin: Rhames and Pepyat.

Beun, R.-J., E. de Vos, and C. Witterman (2003). Embodied Conversational Agents: Effects on Memory Performance and Anthropomorphisation. In: *IVA 2003*. Vol. 2792. Lecture Notes in Artificial Intelligence. Ed. by T. Rist, R. S. Aylett, D. Ballin, and J. Rickel. Berlin/Heidelberg: Springer, 315–319.

Bloor, D. (1996). Idealism and the Sociology of Knowledge. *Social Studies of Science* 26, 839–856.

Bluff, L. A., J. Troscianko, A. A. Weir, A. Kacelnik, and C. Rutz (2010). Tool Use by Wild New Caledonian Crows Corvus Moneduloides at Natural Foraging Sites. *Proceedings of the Royal Society, B* 277, 1377–1385.

Boden, M. A. (2006). *Mind as Machine: A History of Cognitive Science*. Oxford: Oxford University Press.

Boesch, C. (1993). Aspects of Transmission of Tool-Use in Wild Chimpanzees. In: *Tools, Language and Cognition in Human Evolution*. Ed. by K. R. Gibson and T. Ingold. Cambridge: Cambridge University Press, 171–183.

Boesch, C. and H. Boesch (1990). Tool Use and Tool Making in Wild Chimpanzees. *Folia Primatologica* 54, 86–99.

Bogner, A., B. Littig, and W. Menz, eds. (2009). *Interviewing Experts*. Research Methods. Basingstoke: Palgrave Macmillan.

Böhme, G. (2008). *Invasive Technisierung: Technikphilosophie und Technikkritik*. Kusterdingen: Die Graue Edition.

Boumans, M. (2013). Affordance Affords Measurement. Paper presented at "What Affordance Affords" workshop, TU Darmstadt, Germany, on 25 November, 2013.

Boyd, R., P. Gasper, and J. Trout, eds. (1991). *The Philosophy of Science*. Cambridge: MIT Press.

Boyd, R. and P. J. Richerson (1985). *Culture and the Evolutionary Process*. Chicago: University of Chicago Press.

Brandon, R. N. (1995). *Adaptation and Environment*. 2nd ed. Princeton: Princeton University Press.

Breazeal, C. (2002). *Designing Sociable Robots*. Cambridge: MIT Press.

———. (2003). Towards Sociable Robots. *Robotics and Autonomous Systems* 42, 167–175.

Breazeal, C., J. Gray, and M. Berlin (2009). An Embodied Cognition Approach to Mindreading Skills for Socially Intelligent Robots. *The International Journal of Robotics Research* 28.5, 656–680. DOI: 10.1177/0278364909102796.

Brentano, F. (1874). *Psychologie vom empirischen Standpunkte*. Vol. 1–2. Leipzig: Duncker & Humblot.

Brooks, R. (1991). New Approaches to Robotics. *Science* 253, 1227–1232.

———. (1999). *Cambrian Intelligence: The Early History of the New AI*. Cambridge/London: MIT Press.

———. (2002a). *Flesh and Machines*. New York: Pantheon.

———. (2002b). Humanoid Robots. *Communications of the ACM* 45.3, 33–38.

Brown, E. and P. Cairns (2004). A Grounded Investigation of Game Immersion. In: *CHI 2004*. Vienna: ACM Press, 1279–1300.

Burge, T. (1979). Individualism and the Mental. *Midwest Studies in Philosophy* 4.1, 73–121.

Byrne, R. W. and A. Whiten, eds. (1988). *Machiavellian Intelligence: Social Expertise and the Evolution of Intellect in Monkeys, Apes and Humans*. Oxford: Oxford University Press.

Carnap, R. (1928). *Der Logische Aufbau der Welt*. Leipzig: Felix Meiner Verlag.

———. (1931). Überwindung der Metaphysik durch logische Analyse der Sprache. *Erkenntnis* 2, 219–241.

———. (1955). Logical Foundations of the Unity of Science. In: *International Encyclopedia of Unified Science*. Ed. by O. Neurath, R. Carnap, and C. W. Morris. Vol. I. Chicago: University of Chicago Press, 32–62.

———. (1959). The Elimination of Metaphysics through Logical Analysis of Language. In: *Logical Positivism*. Ed. by A. J. Ayer. New York: Free Press, 60–81.

Caro, M. de and D. Macarthur (2004). Introduction: The Nature of Naturalism. In: *Naturalism in Question*. Ed. by M. de Caro and D. Macarthur. Cambridge: Harvard University Press, 1–18.

Carter, J. A., J. Kallestrup, S. O. Palermos, and D. Pritchard (2014). Varieties of Externalism. Philosophical *Issues* 24, 63–109.

Cassirer, E. (1944). *An Essay on Man: An Introduction to a Philosophy of Human Culture*. Yale/New Haven: Yale University Press.

Cavalli-Sforza, L. L. and M. W. Feldman (1981). *Cultural Transmission and Evolution: A Quantitative Approach*. Princeton: Princeton University Press.

Chaminade, T. and G. Cheng (2009). Social Cognitive Neuroscience and Humanoid Robotics. Journal *of Physiology – Paris* 103, 286–295.

Chemero, A. (2003a). An Outline of a Theory of Affordances. *Ecological Psychology* 15.2, 181–195.

———. (2003b). Information for Perception and Information Processing. *Minds and Machines* 13, 577–588.

———. (2009). *Radical Embodied Cognitive Science*. Cambridge/London: MIT Press.

Cheng, G. (2014). *Humanoid Robotics and Neuroscience: Science, Engineering and Society*. Frontiers in Neuroengineering Series. Boca Raton: CRC Press.

Chesher, C. (2004). Neither Gaze nor Glance, but Glaze: Relating to Console Game Screens. *Scan Journal* 1.1, html. URL: http://scan.net.au/scan/journal/display. php?journal_id=19 (visited on 02/22/2013).

Chomsky, N. (2006). *Language and Mind*. 3rd ed. Cambridge/New York: Cambridge University Press.

Christiansen, M. H. and S. Kirby, eds. (2003). *The Evolution of Language*. Oxford/New York: Oxford University Press.

Clark, A. (1997). *Being There: Putting Brain, Body and World Together Again*. Cambridge: MIT Press.

———. (1998). Magic Words: How Language Augments Human Computation. In: *Language and Thought: Interdisciplinary Themes*. Ed. by P. Carruthers and J. Boucher. Cambridge: Cambridge University Press, 162–183.

———. (2003). *Natural-Born Cyborgs: Minds, Technologies, and the Future of Human Intelligence*. Oxford/New York: Oxford University Press.

———. (2006). Language, Embodiment, and the Cognitive Niche. *Trends in Cognitive Sciences* 10.8, 370–374.

———. (2007). Mikrofunktionalismus: Konnektionismus und die wissenschaftliche Erklärung mentaler Zustände. In: *Grundkurs Philosophie des Geistes*. Ed. by T. Metzinger. Vol. 2. Paderborn: Mentis, 392–426.

———. (2008). *Supersizing the Mind: Embodiment, Action, and Cognitive Extension*. Oxford/New York: Oxford University Press.

———. (2010a). Coupling, Constitution, and the Cognitive Kind: A Reply to Adams and Aizawa. In: *The Extended Mind.* Ed. by R. Menary. Cambridge/London: MIT Press, 81–99.

———. (2010b). Memento's Revenge: The Extended Mind, Extended. In: *The Extended Mind.* Ed. by R. Menary. Cambridge/London: MIT Press, 43–66.

Clark, A. and D. Chalmers (1998). The Extended Mind. *Analysis* 58.1, 7–19.

Clark, A. and J. Toribio (1994). Doing without Representing? *Synthese* 101, 401–431.

Cole, C. (2013). *Continental's "Driver Focus Vehicle" Showcases Future Tech to Curb Distracted Driving.* URL: http://www.autoguide.com/auto-news/tag/driver-focus-vehicle (visited on 02/07/2013).

Collins, H. M. and M. Kusch (1998). *The Shape of Actions: What Humans and Machines Can Do.* Cambridge: MIT Press.

Copeland, B. J. (2000). The Turing Test. *Minds and Machines* 10, 519–539.

———. (2009). The Church-Turing Thesis. In: *The Stanford Encyclopedia of Philosophy.* Ed. by E. N. Zalta. Spring. Stanford: The Metaphysics Research Lab, html. URL: http://plato.stanford.edu/archives/spring2009/entries/church-turing/.

Costall, A. (2004). From Direct Perception to the Primacy of Action: A Closer Look at James Gibson's Ecological Approach to Psychology. In: *Theories of Infant Development.* Ed. by G. Bremner and A. Slater. Oxford: Blackwell, 70–89.

Crary, A. and R. Read, eds. (2000). *The New Wittgenstein.* London/New York: Routledge.

Cross, K. (2016). *Augmented Reality Games Like Pokémon Go Need a Code of Ethics – Now.* URL: https://www.wired.com/2016/08/ethics-ar-pokemon-go/ (visited on 10/01/2016).

da Costa, N. and S. French (2003). *Science and Partial Truth: A Unitary Approach to Models and Scientific Reasoning.* Oxford/New York: Oxford University Press.

Darwin, C. (1859). *On The Origin of Species by Means of Natural Selection: Or the Preservation of Favoured Races in the Struggle for Life.* 1st ed. London: John Murray.

Davidson, D. (1980). *Essays on Actions and Events.* Oxford: Oxford University Press.

Dawkins, R. (1983). Universal Darwinism. In: *Evolution from Molecules to Men.* Ed. by D. S. Bendall. Cambridge/London: Cambridge University Press, 403–425.

———. (1999). *The Extended Phenotype: The Long Reach of the Gene.* Revised edition with new afterword and further reading; originally published 1982. Oxford/New York: Oxford University Press.

Deacon, T. W. (1997). *The Symbolic Species: The Co-Evolution of Language and the Brain.* New York/London: Norton.

Dennett, D. C. (1987). *The Intentional Stance.* Cambridge/London: MIT Press.

———. (1995). *Darwin's Dangerous Idea.* London: Penguin.

———. (1996). Granny versus Mother Nature – No Contest. *Mind and Language,* 11, 263–269.

DeSteno, D., C. Breazeal, R. H. Frank, D. Pizarro, J. Baumann, L. Dickens, and J. J. Lee (2012). Detecting the Trustworthiness of Novel Partners in Economic Exchange. *Psychological Science* 23.12, 1549–1556. DOI: 10.1177/0956797612448793.

Dewey, J. (1910). The Influence of Darwinism on Philosophy. In: *The Influence of Darwinism on Philosophy and Other Essays in Contemporary Thought.* New York: Holt, 1–19.

———. (1929). *Experience and Nature.* London: George Allen & Unwin.

Dieber, B., T. Grassauer, J. Mayring, and B. Rinner (2010). The Geobashing Architecture for Location-Based Mobile Massive Multiplayer Online Games. Working paper, Institute of Networked and Embedded Systems, University of Klagenfurt. Klagenfurt.

Dijk, L. van, R. Withagen, and R. M. Bongers (2015). Information without Content: A Gibsonian Reply to Enactivists' Worries. *Cognition* 134, 210–214.

Di Paolo, E. A. (2003). Evolving Spike-Timing-Dependent Plasticity for Single-Trial Learning in Robots. *Philosophical Transactions of the Royal Society of London. Series A: Mathematical, Physical and Engineering Sciences* 361.1811, 2299–2319. DOI: 10.1098/rsta.2003.1256.

Doncieux, S. and J.-B. Mouret (2014). Beyond Black-Box Optimization: A Review of Selective Pressures for Evolutionary Robotics. *Evolutionary Intelligence* 7.2, 71–93. DOI: 10.1007/ s12065–014–0110-x.

Dretske, F. (1981). *Knowledge and the Flow of Information*. Cambridge: MIT Press.

———. (1983). Précis of Knowledge and the Flow of Information. *Behavioral and Brain Sciences* 6, 55–63.

———. (1985). Machines and the Mental. *Proceedings and Addresses of the American Philosophical Association* 59.1, 23–33.

———. (1986). Misrepresentation. In: *Belief: Form, Content, and Function*. Ed. by R. J. Bogdan. Oxford: Clarendon Press, 17–36.

———. (1988). *Explaining Behavior: Reasons in a World of Causes*. Cambridge/London: MIT Press.

———. (2009). Information-Theoretic Semantics. In: *The Oxford Handbook of Philosophy of Mind*. Ed. by A. Beckermann, B. P. McLaughlin, and S. Walter. Oxford: Oxford University Press, 381–393.

Dreyfus, H. L. (1979). *What Computers Can't Do: A Critique of Artificial Reason*. New York/London: Harper & Row.

Ducheneaut, N., M.-H. Wen, N. Yee, and G. Wadley (2009). Body and Mind: A Study of Avatar Personalization in Three Virtual Worlds. In: *CHI'09: New Media Experiences*. New York: ACM, 1151–1160.

Economist (2016). *I Mug you, Pikachu!* URL: http://www.economist.com/news/business-and-finance/21702087-nintendo-shares-rocket-after-its-successful-foray-smartphone-gaming-pok-mon-go-shows (visited on 10/01/2016).

Edwards, P. N. (1994). From "Impact" to Social Process: Computers in Society and Culture. In: *Handbook of Science and Technology Studies*. Ed. by Sheila Jasanoff, G. Markle, J. Petersen, and T. Pinch. Beverly Hills: Sage, 257–285.

Einstein, A. (1918). Motive des Forschens. In: *Zu Max Plancks sechzigstem Geburtstag. Ansprachen, gehalten am 26. April 1918 in der Deutschen Physikalischen Gesellschaft, Berlin*. Ed. by E. Warburg and M. Planck. Karlsruhe: C. F. Müllersche Hofbuchhandlung, 29–32.

Eldredge, N. and S. J. Gould (1972). Punctuated Equilibria: An Alternative to Phyletic Gradualism. In: *Models in Paleobiology*. Ed. by T. Schopf. San Francisco: Freeman & Co., 82–115.

Elton, C. S. (1927). *Animal Ecology*. London: Sidgwick and Jackson.

Evans, R. and H. M. Collins (2010). Interactional Expertise and the Imitation Game. In: *Trading Zones and Interactional Expertise: Creating New Kinds of Collaboration*. Ed. by M. E. Gorman. Cambridge: MIT Press, 53–70.

Farina, M. (2013). Neither Touch nor Vision: Sensory Substitution as Artificial Synaesthesia? Biology *& Philosophy* 28, 639–655.

———. (2016). Three Approaches to Human Cognitive Development: Neo-Nativism, Neuroconstructivism, and Dynamic Enskillment. *British Journal for the Philosophy of Science* 67, 617–641.

Feiner, S. K. (2002). Augmented Reality: A New Way of Seeing. *Scientific American* 286.4, 48–55.

Feyerabend, P. K. (1975). *Against Method: Outline of an Anarchist Theory of Knowledge*. London/New York: New Left Books.

Floridi, L. (2004). Outline of a Theory of Strongly Semantic Information. *Minds and Machines* 14, 197–221.

———. (2011). Semantic Conceptions of Information. In: *The Stanford Encyclopedia of Philosophy*. Ed. by E. N. Zalta. Spring. Stanford: The Metaphysics Research Lab, html. URL: http://plato.stanford.edu/entries/information-semantic/.

Fodor, J. (1981). The Mind-Body Problem. *Scientific American* 244.1, 114–123.

———. (1990). *A Theory of Content and Other Essays*. Cambridge: MIT Press.

———. (1996). Deconstructing Dennett's Darwin. *Mind and Language* 11, 246–262.

Fodor, J. and M. Piattelli-Palmarini (2010). *What Darwin Got Wrong*. New York: Farrar, Straus, & Giroux.

Fodor, J. and Z. Pylyshyn (1981). How Direct Is Visual Perception? Some Reflections on Gibson's 'Ecological Approach'. *Cognition* 9, 139–196.

Frege, G. (1884). *Die Grundlagen der Arithmetik: Eine logisch mathematische Untersuchung über den Begriff der Zahl*. Breslau: Wilhelm Koebner.

Gardner, H. (1985). *The Mind's New Science: A History of the Cognitive Revolution*. New York: Basic Books.

Gäser, J. and G. Laudel (2009). *Experteninterviews und qualitative Inhaltsanalyse*. 4th ed. Wiesbaden: VS Verlag für Sozialwissenschaften.

Gehlen, A. (1940). *Der Mensch. Seine Natur und seine Stellung in der Welt*. Berlin: Junker und Dünnhaupt.

Gibson, E. J. and R. D. Walk (1960). The 'Visual Cliff'. *Scientific American* 202.4, 64–71.

Gibson, J. J. (1960). The Information Contained in Light. *Acta Psychologica* 17, 23–30.

———. (1966). *The Senses Considered as Perceptual Systems*. Boston: Houghton Mifflin.

———. (1970). On the Relation between Hallucination and Perception. *Leonardo* 3.4, 425–427.

———. (1971). The Information Available in Pictures. *Leonardo* 4, 27–35.

———. (1973). On the Concept of 'Formless Invariants' in Visual Perception. *Leonardo* 6.1, 43–45.

———. (1979). *The Ecological Approach to Visual Perception*. Reprint ed. 1986. New York/Hove: The Psychology Press. Boston: Houghton Mifflin.

Gibson, K. R. and T. Ingold, eds. (1993). *Tools, Language and Cognition in Human Evolution*. Cambridge: Cambridge University Press.

Giere, R. N. (2009). Naturalism. In: *A Companion to the Philosophy of Science*. Ed. by W. Newton-Smith. Oxford: Blackwell, 308–310.

Godfrey-Smith, P. (1994). A Modern History Theory of Functions. *Noûs* 28, 344–362.

———. (1996a). *Complexity and the Function of Mind in Nature*. Cambridge: Cambridge University Press.

———. (1996b). Spencer and Dewey on Life and Mind. In: *The Philosophy of Artificial Life*. Ed. by M. A. Boden. Oxford: Oxford University Press, 314–331.

———. (2013a). Information and Influence in Sender-Receiver Models, with Applications to Animal Behavior. In: *Animal Communication Theory: Information and Influence*. Ed. by U. Stegmann. Cambridge: Cambridge University Press, 377–396.

———. (2013b). Sender-Receiver Systems within and between Organisms. In: Philosophy of Science Association: 23rd Biennial Meeting, San Diego, CA, 2012. URL: http://philsci-archive.pitt.edu/9513/.

Goldschmidt, R. B. (1949). Phenocopies. *Scientific American* 181.4, 46–49.

Gombrich, E., J. J. Gibson, and R. Arnheim (1971). Exchange of Letters. *Leonardo* 4, 195–203.

Goodman, N. (1976). *Languages of Art: An Approach to a Theory of Symbol*. 2nd ed. Indianapolis: Hackett.

Gordon, D. M. (2001). The Development of Ant Colony Behavior. In: *Cycles of Contingency: Developmental Systems and Evolution*. Ed. by S. Oyama, P. E. Griffiths, and R. D. Gray. Cambridge/London: MIT Press, 141–148.

Gould, S. J. and R. C. Lewontin (1979). The Spandrels of San Marco and the Panglossian Paradigm: A Critique of the Adaptationist Programme. *Proceedings of the Royal Society, B* 205, 581–598.

Gould, S. J. and E. S. Vrba (1982). Exaptation – A Missing Term in the Science of Form. *Paleobiology* 8, 4–15.

Gray, J. and C. Breazeal (2014). Manipulating Mental States through Physical Action. *International Journal of Social Robotics* 6.3, 315–327. ISSN: 1875–4805. DOI: 10.1007/s12369-014-0234-2.

Griffiths, P. E. and R. D. Gray (2001). Darwinism and Developmental Systems. In: *Cycles of Contingency: Developmental Systems and Evolution*. Ed. by S. Oyama, P. E. Griffiths, and R. D. Gray. Cambridge/London: MIT Press, 195–218.

Grinnell, J. (1917). The Niche Relationship of the California Thrasher. *Auk* 34, 427–433.

Gunderson, K. (1964). The Imitation Game. *Mind* 73, 234–245.

Harman, G. (1983). Knowledge and the Relativity of Information. *Behavioral and Brain Sciences* 6, 72.

Harvey, I., P. Husbands, and D. Cliff (1994). Seeing the Light: Artificial Evolution, Real Vision. In: *From Animals to Animats 3: Proceedings of the Third International Conference on Simulation of Adaptive Behavior*. Ed. by D. Cliff, P. Husbands, J.-A. Meyer, and R. W. Wilson. Cambridge/London: MIT Press, 392–401.

Head, H. (1920). *Studies in Neurology*. London: Oxford University Press.

Heft, H. (2001). *Ecological Psychology in Context: James Gibson, Roger Barker, and the Legacy of William James's Radical Empiricism*. Mahwah/London: Lawrence Erlbaum Associates.

Herrmann, E., J. Call, M. V. Hernández-Lloreda, B. Hare, and M. Tomasello (2007). Humans Have Evolved Specialized Skills of Social Cognition: The Cultural Intelligence Hypothesis. *Science* 317, 1360–1366.

Hjorth, L. (2011). Mobile@Game Cultures: The Place of Urban Mobile Gaming. *Convergence* 17.4, 357–371.

Hjorth, L. and I. Richardson (2011). Playing the Waiting Game: Complicating Notions of (Tele)presence and Gendered Distraction in Casual Mobile Gaming. In: *Cultures of Participation: Media Practices, Politics and Literacy*. Ed. by H. Greif, L. Hjorth, A. Lasén, and C. Lobet-Maris. Berlin: Peter Lang, 111–125.

Ho, C.-C. and K. F. MacDorman (2010). Revisiting the Uncanny Valley Theory: Developing and Validating an Alternative to the Godspeed Indices. *Computers in Human Behavior* 26.6, 1508–1518.

Hodges, A. (1983). *Alan Turing: The Enigma*. New York: Simon & Schuster.

Humphrey, N. (1976). The Social Function of Intellect. In: *Growing Points in Ethology*. Ed. by P. Bateson and R. Hinde. Cambridge: Cambridge University Press, 303–317.

Hurley, S. L. (1998). Vehicles, Contents, Conceptual Structure, and Externalism. *Analysis* 58.1, 1–6.

———. (2010). The Varieties of Externalism. In: *The Extended Mind*. Ed. by R. Menary. Cambridge/London: MIT Press, 101–153.

Husbands, P. and O. Holland (2008). The Ratio Club: A Hub of British Cybernetics. In: *The Mechanical Mind in History*. Ed. by P. Husbands, O. Holland, and M. Wheeler. Cambridge/London: MIT Press, 91–148.

Hussain, Z. and M. D. Griffiths (2008). Gender Swapping and Socializing in Cyberspace: An Exploratory Study. *Cyberpsychology and Behavior* 11, 47–53.

Hutchins, E. (2005). Material Anchors for Conceptual Blends. *Journal of Pragmatics* 37, 1555–1577.

Hutchinson, G. E. (1957). Concluding Remarks. *Cold Spring Harbor Symposia on Quantitative Biology* 22, 415–427. DOI: 10.1101/SQB.1957.022.01.039.

Jennett, C., A. L. Cox, P. Cairns, S. Dhoparee, A. Epps, T. Tijs, and A. Walton. (2008). Measuring and Defining the Experience of Immersion in Games. *International Journal of Human-Computer Studies* 66.9, 641–661. DOI: 10.1016/j.ijhcs.2008.04.004.

Juul, J. (2010). *A Casual Revolution: Reinventing Video Games and Their Players*. Cambridge: MIT Press.

Kant, I. (1781). *Kritik der reinen Vernunft*. Riga: Johann Friedrich Hartknoch.

Keil, G. (2008). Naturalism. In: *The Routledge Companion to Twentieth Century Philosophy*. Ed. by D. Moran. London: Routledge, 254–307.

Kennedy, J. M., C. D. Green, A. Nicholls, and C. H. Liu (1992). Illusions and Knowing What Is Real. *Ecological Psychology* 4.3, 153–172.

King Face (1987). *King Face*. 12" EP. Self-released.

Kiverstein, J. and M. Farina (2011). Embraining Culture: Leaky Minds and Spongy Brains. Theorema 30.2, 35–53.

Kiverstein, J., M. Farina, and A. Clark (2013). *The Extended Mind Thesis*. URL: http://www.oxfordbibliographies.com/view/document/obo-9780195396577/obo-9780195396577–0099.xml (visited on 11/27/2013).

———. (2015). Substituting the Senses. In: *The Oxford Handbook of the Philosophy of Perception*. Ed. by M. Matthen. Oxford: Oxford University Press.

Knight, W. (2016a). *If a Driverless Car Goes Bad We May Never Know Why*. URL: https://www.technologyreview.com/s/601860/if-a-driverless-car-goes-bad-we-may-never-know-why/ (visited on 07/07/2016).

———. (2016b). *Tesla Crash Will Shape the Future of Automated Cars*. URL: https://www.technologyreview.com/s/601829/tesla- crash- will- shape- the- future-of-automated-cars/ (visited on 07/01/2016).

Kohler, C. (2016). *Pokemon Go Will Soon Bring Monsters into Your (Real) World*. URL: https://www.wired.com/2016/06/pokemon-go-preview/ (visited on 01/10/2916).

Korhonen, H., H. Saarenpää, and J. Paavilainen (2008). Pervasive Mobile Games – A New Mindset for Players and Developers. In: *Fun and Games*. Vol. 5294. Lecture Notes in Computer Science. Ed. by P. Markopoulos, B. de Ruyter, W. IJsselsteijn, and D. Rowland. Berlin/Heidelberg: Springer, 21–32.

Krummheuer, A. L. (2008). Die Herausforderung künstlicher Handlungsträgerschaft. Frotzelattacken in hybriden Austauschprozessen von Menschen und virtuellen Agenten. In: *Information und Gesellschaft. Technologien einer sozialen Beziehung*. Ed. by H. Greif, O. Mitrea, and M. Werner. Wiesbaden: VS Research, 73–95.

———. (2009). Conversation Analysis, Video Recordings and Human-Computer Interchanges. In: *Video Interaction Analysis: Methods and Methodology*. Ed. by U. T. Kissmann. Berlin: Peter Lang, 59–83.

Lee, J. J., W. B. Knox, J. B. Wormwood, C. Breazeal, and D. DeSteno (2013). Computationally Modeling Interpersonal Trust. *Frontiers in Psychology* 4, article 893.

Leibniz, G. W. (1685). *Machina arithmetica in qua non aditio tantum et subtractio sed et multiplicatio nullo, divisio vero paene nullo animi labore peragantur*. Manuscript.

Lettvin, J. Y., H. R. Maturana, W. S. McCulloch, and W. H. Pitts (1968). What the Frog's Eye Tells the Frog's Brain. In: *The Mind: Biological Approaches to Its Functions*. Ed. by W. C. Corning and M. Balaban. New York: Interscience, 233–258.

Lewis, D. (1969). *Convention*. Cambridge: Harvard University Press.

———. (1980). Mad Pain and Martian Pain. In: *Readings in Philosophy of Psychology*. Ed. by N. Block. Vol. 1. Cambridge: Harvard University Press, 216–222.

Lewontin, R. C. (1982). Organism and Environment. In: *Learning, Development, and Culture*. Ed. by H. C. Plotkin. Chichester: Wiley & Sons, 151–170.

———. (2000). *The Triple Helix: Gene, Organism, and Environment*. Cambridge/London: Harvard University Press.

Lin, A., D. Schwarz, R. Sellaouti, S. Shokur, R. C. Moioli, F. L. Brasil, K. R. Fast, N. A. Peretti, A. Takigami, S. Gallo, K. Lyons, P. Mittendorfer, M. Lebedev, S. Joshi, G. Cheng, E. Morya, A. Rudolph, and M. Nicolelis. (2014). The Walk Again Project: Brain-Controlled Exoskeleton Locomotion. Poster presented at Neuroscience *Meeting 2014*. Washington, DC: Society for Neuroscience.

Lobo, S. (2016). *Smartphone vorm Gesicht: Was ist das für eine komische Geste beim Telefonieren?* URL: http://www.spiegel.de/netzwelt/gadgets/smartphone-vor-dem-mund-was-ist-das-fuer-eine-komische-geste-a-1111245.html (visited on 09/07/2016).

Macdonald, G. and D. Papineau, eds. (2006). *Teleosemantics*. Oxford: Oxford University Press.

Maes, P. (1993). Behavior-Based Artificial Intelligence. In: *From Animals to Animats 2: Proceedings of the Second International Conference on Simulation of Adaptive Behavior*. Ed. by J.-A. Meyer and S. W. Wilson. Cambridge: MIT Press, 2–10.

Maes, P. and P. Mistry (2009). *Unveiling the "Sixth Sense," Game-Changing Wearable Tech*. URL: http://www.ted.com/talks/pattie_maes_demos_the_sixth_sense (visited on 06/30/2016).

Marcus, E. (2005). Mental Causation in a Physical World. *Philosophical Studies* 122, 27–50.

Marr, D. (2010). *Vision: A Computational Investigation into the Human Representation and Processing of Visual Information*. Originally published 1982. Cambridge: MIT Press.

Matsuzawa, T., ed. (2001). *Primate Origins of Human Cognition and Behavior*. Tokyo/Berlin: Springer.

Maynard Smith, J. (2000). The Concept of Information in Biology. *Philosophy of Science* 67, 177–194.

McCorduck, P. (2004). *Machines Who Think: A Personal Inquiry into the History and Prospects of Artificial Intelligence*. 2nd ed. Natick: AK Peters.

McGinn, C. (1977). Charity, Interpretation, and Belief. *Journal of Philosophy* 74.9, 521–535.

McGrew, W. (2013). Is Primate Tool Use Special? Chimpanzee and New Caledonian Crow Compared. *Philosophical Transactions of the Royal Society, B* 368, 20120422.

Menary, R. (2010a). Cognitive Integration and the Extended Mind. In: *The Extended Mind*. Ed. by R. Menary. Cambridge/London: MIT Press, 227–243.

———. (2010b). Introduction: The Extended Mind in Focus. In: *The Extended Mind*. Ed. by R. Menary. Cambridge/London: MIT Press, 1–25.

———. (2010c). Introduction to the Special Issue on 4E Cognition. *Phenomenology and the Cognitive Sciences* 9.4. Special Issue "4E Cognition: Embodied, Embedded, Enacted, Extended", 459–463.

———. ed. (2010d). *The Extended Mind*. Cambridge/London: MIT Press.

Mennecke, B. E., D. McNeil, E. M. Roche, D. R. Bray, A. M. Townsend, and J. Lester (2008). Second Life and Other Virtual Worlds: A Roadmap for Research. *Communications of the Association for Information Systems* 22, 371–388.

Meuser, M. and U. Nagel (2009). Das Experteninterview – konzeptionelle Grundlagen und methodische Anlage. In: *Methoden der vergleichenden Politik- und Sozialwissenschaft*. Ed. by S. Pickel, G. Pickel, H.-J. Lauth, and D. Jahn. Wiesbaden: VS Verlag für Sozialwissenschaften, 465–479.

Milgram, P. and F. Kishino (1994). A Taxonomy of Mixed Reality Visual Displays. *IEICE Transactions on Information Systems* E77-D.12, 1321–1329.

Millikan, R. G. (1984). *Language, Thought, and Other Biological Categories*. Cambridge/London: MIT Press.

———. (1986). Thoughts without Laws; Cognitive Science with Content. *Philosophical Review* 95, 47–80.

———. (1993a). Explanation in Biopsychology. In: *Mental Causation*. Ed. by J. Heil and A. Mele. Oxford: Oxford University Press, 211–232.

———. (1993b). *White Queen Psychology and Other Essays for Alice*. Cambridge/London: MIT Press.

———. (1995). Pushmi-Pullyu Representations. *Philosophical Perspectives* 9, 185–200.

———. (2000). *On Clear and Confused Ideas: An Essay about Substance Concepts*. Cambridge/New York: Cambridge University Press.

———. (2001). What Has Natural Information to do with Intentional Representation? In: *Naturalism, Evolution and Mind*. Ed. by D. M. Walsh. Cambridge: Cambridge University Press, 105–125.

———. (2004). *Varieties of Meaning*. Cambridge/London: MIT Press.

———. (2007). Reply to Recanati. *Philosophy and Phenomenological Research* 75.3, 682–691.

———. (2017). *Beyond Concepts: Unicepts, Language and Information*. Oxford: Oxford University Press.

Mistry, P., P. Maes, and L. Chang (2009). WUW – Wear Ur World: A Wearable Gestural Interface. In: *Proceedings of the 27th International Conference on Human Factors in Computing Systems – CHI EA '09*. Boston: Association for Computing Machinery/Special Interest Group on Computer-Human Interaction (ACM SIGCHI), 4111–4116.

Mitcham, C. (1994). *Thinking through Technology: The Path between Engineering and Philosophy*. Chicago: University of Chicago Press.

Mittendorfer, P. and G. Cheng (2011). Self-Organizing Sensory-Motor Map for Low-Level Touch Reactions. en. In: *11th IEEE-RAS International Conference on Humanoid Robots*. New York: IEEE, 59–66.

Montola, M., J. Stenros, and A. Waern, eds. (2009). *Pervasive Games: Theory and Design*. Burlington: Morgan Kaufmann Publishers.

Moor, J. H. (1976). An Analysis of the Turing Test. *Philosophical Studies* 30, 249–257.

Mori, M. (1970). Bukimi no tani (The Uncanny Valley). *Energy* 7.4, 33–35. URL: http://www.androidscience.com/theuncannyvalley/proceedings2005/uncannyvalley.html (visited on 07/06/2010).

Mumford, L. (1934). *Technics and Civilization*. New York: Harcourt, Brace & World.

Nagel, E. (1961). *The Structure of Science*. New York: Harcourt, Brace & World. Chap. 12: Mechanistic Explanation and Organismic Biology.

National Transportation Safety Board (2016). *Preliminary Report, Highway HWY16FH018*. URL: http://www.ntsb.gov/investigations/AccidentReports/Pages/HWY16FH018-preliminary.aspx (visited on 09/29/2016).

Neander, K. (1991a). Functions as Selected Effects: The Conceptual Analyst's Defense. *Philosophy of Science* 58.2, 168–184.

———. (1991b). The Teleological Notion of 'Function'. *Australasian Journal of Philosophy* 69.1, 454–468.

Nelson, A. L. (2014). Embodied Artificial Life at an Impasse: Can Evolutionary Robotics Methods Be Scaled? In: *Evolving and Autonomous Learning Systems (EALS), 2014 IEEE Symposium on*, 25–34. DOI: 10.1109/EALS.2014.7009500.

Neurath, O. (1983). Protocol Statements. In: *Philosophical Papers 1913–1946*. Ed. by R. Cohen and M. Neurath. Dordrecht: Reidel, 91–99.

Nicolelis, M. A. L. and J. K. Chapin (2002). Controlling Robots with the Mind. *Scientific American* 287.4, 46–53. URL: http://www.scientificamerican.com/article/controlling-robots-with-the-mind/.

Nolfi, S. and D. Floreano (2000). *Evolutionary Robotics: The Biology, Intelligence and Technology of Self-Organizing Machines*. Cambridge/London: MIT Press.

Norman, D. A. (2002). *The Design of Everyday Things*. New York: Basic Books.

Odling-Smee, F. J., K. N. Laland, and M. W. Feldman (1996). Niche Construction. *The American Naturalist* 147.4, 641–648.

———. (2003). *Niche Construction: The Neglected Process in Evolution*. Princeton: Princeton University Press.

Oppenheim, P. and H. Putnam (1958). Unity of Science as a Working Hypothesis. In: *Minnesota Studies in the Philosophy of Science*. Ed. by H. Feigl, M. Scriven, and G. Maxwell. Vol. II. Minneapolis: University of Minnesota Press, 3–36.

Owen, R. (1848). *On the Archetype and Homologies of the Vertebrate Skeleton*. London: John van Voorst.

Oyama, S. (2000). *The Ontogeny of Information: Developmental Systems and Evolution*. 2nd ed. Durham: Duke University Press.

Oyama, S., P. E. Griffiths, and R. D. Gray, eds. (2001). *Cycles of Contingency: Developmental Systems and Evolution*. Cambridge/London: MIT Press.

Papineau, D. (1987). *Reality and Representation*. Oxford: Blackwell.

———. (1993). *Philosophical Naturalism*. Oxford: Blackwell.

———. (2007). Naturalism. In: *The Stanford Encyclopedia of Philosophy*. Ed. by E. N. Zalta. Stanford: The Metaphysics Research Lab, html. URL: http://plato.stanford.edu/archives/spr2009/entries/naturalism/.

Parikka, J. and J. Souminen (2006). Victorian Snakes? Towards a Cultural History of Mobile Games and the Experience of Movement. *Game Studies: The International Journal of Computer Game Research* 6.

Pearce, T. (2014). The Origins and Development of the Idea of Organism-Environment Interaction. In: *Entangled Life: Organism and Environment in the Biological and Social Sciences*. Ed. by G. Barker, E. Desjardins, and T. Pearce. Dordrecht: Springer, 13–32.

Peirce, C. S. (1868). On a New List of Categories. *Proceedings of the American Academy of Arts and Sciences* 7, 287–298. URL: http://www.cspeirce.com/menu/library/bycsp/newlist/newlist.htm.

———. (1998). What Is a Sign? In: *The Essential Peirce: Selected Philosophical Writings*. Ed. by N. Houser and C. Kloesel. Vol. 2. Manuscript written c. 1894. Bloomington: Indiana University Press, 4–10. URL: http://www.iupui.edu/~peirce/ep/ep2/ep2book/ch02/ep2ch2.htm.

Peterson, A. (2016). *Holocaust Museum to Visitors: Please Stop Catching Pokémon Here*. URL: https://www.washingtonpost.com/news/the-switch/wp/2016/07/12/holocaust-museum-to-visitors-please-stop-catching-pokemon-here/ (visited on 10/01/2016).

Pinch, T. J. and W. E. Bijker (1984). The Social Construction of Facts and Artefacts: Or How the Sociology of Science and the Sociology of Technology Might Benefit Each Other. *Social Studies of Science* 14.3, 399–441.

Pinker, S. and P. Bloom (1990). Natural Language and Natural Selection. *Behavioral and Brain Sciences* 13, 707–784.

Pittendrigh, C. S. (1958). Adaptation, Natural Selection, and Behavior. In: *Behaviour and Evolution*. Ed. by A. Roe and G. G. Simpson. New Haven: Yale University Press, 390–416.

Plessner, H. (1928). *Die Stufen des Organischen und der Mensch. Einleitung in die philosophische Anthropologie*. Berlin/Leipzig: de Gruyter.
Popper, K. R. (1959). *The Logic of Scientific Discovery*. London: Hutchinson.
Price, C. S. (2001). *Functions in Mind: A Theory of Intentional Content*. Oxford: Oxford University Press.
Proust, J. (1995). Functionalism and Multirealizability: On Interaction between Structure and Function. In: *Science, Mind and Art*. Ed. by K. Gavroglu, J. Stachel, and M. W. Wartofsky. Dordrecht: Springer Netherlands, 169–185.
Purves, D., R. B. Lotto, S. M. Williams, and Z. Xang (2001). Why We See Things the Way We Do: Evidence for a Wholly Empirical Strategy of Vision. *Philosophical Transactions of the Royal Society of London, B* 356, 285–297.
Purves, D., W. T. Wojtach, and R. B. Lotto (2011). Understanding Vision in Wholly Empirical Terms. *Proceedings of the National Academy of Sciences of the United States of America* 108.Supplement 3, 15588–15595.
Putnam, H. (1975a). *Mind, Language and Reality. Philosophical Papers, Vol. 2*. Cambridge: Cambridge University Press.
———. (1975b). The Meaning of 'Meaning'. *Minnesota Studies in the Philosophy of Science* 7, 131–193.
———. (2004). The Content and Appeal of Naturalism. In: *Naturalism in Question*. Ed. by M. de Caro and D. Macarthur. Cambridge: Harvard University Press, 59–70.
Pylyshyn, Z. (1980). Computation and Cognition: Issues in the Foundations of Cognitive Science. *The Behavioral and Brain Sciences* 3, 111–169.
Quine, W. V. O. (1957). The Scope and Language of Science. *British Journal for the Philosophy of Science* 8, 1–17.
———. (1961). Two Dogmas of Empiricism. In: *From a Logical Point of View*. 2nd ed. Cambridge/London: Harvard University Press, 20–46.
———. (1969a). Epistemology Naturalized. In: *Ontological Relativity and other Essays*. New York: Columbia University Press, 69–90.
———. (1969b). Natural Kinds. In: *Ontological Relativity and other Essays*. New York: Columbia University Press, 114–138.
———. (1969c). Ontological Relativity. In: *Ontological Relativity and other Essays*. New York: Columbia University Press, 26–68.
Ramirez-Amaro, K., M. Beetz, and G. Cheng (2015). Transferring Skills to Humanoid Robots by Extracting Semantic Representations from Observations of Human Activities. *Artificial Intelligence* (corrected proof available online from publisher).
Rashid, O., I. Mullins, P. Coulton, and R. Edwards (2006). Extending Cyberspace: Location Based Games Using Cellular Phones. *ACM Computers in Entertainment* 4.1, 3C.
Reed, E. S. (1988). *James J. Gibson and the Psychology of Perception*. New Haven/London: Yale University Press.
Reuters (2016). *DVD Player Found in Tesla Car in Fatal May Crash*. URL: http://www.reuters.com/article/us-tesla-autopilot-dvd-idUSKCN0ZH5BW (visited on 07/01/2016).
Rey, G. (2015). The Analytic/Synthetic Distinction. In: *The Stanford Encyclopedia of Philosophy*. Ed. by E. N. Zalta. Winter. Stanford: The Metaphysics Research Lab, html. URL: http://plato.stanford.edu/archives/win2015/entries/analytic-synthetic/.
Rowlands, M. (2004). *The Body in Mind: Understanding Cognitive Processes*. Cambridge: Cambridge University Press.
———. (2010). *The New Science of the Mind: From Extended Mind to Embodied Phenomenology*. Cambridge/London: MIT Press.

Runeson, S. (1988). The Distorted Room Illusion, Equivalent Configurations, and the Specificity of Static Optic Arrays. *Journal of Experimental Psychology: Human Perception and Performance* 14.2, 295–304.

Rupert, R. D. (2004). Challenges to the Hypothesis of Extended Cognition. *Journal of Philosophy* 101.8, 389–428.

———. (2010). Representation in Extended Cognitive Systems: Does the Scaffolding of Language Extend the Mind? In: *The Extended Mind*. Ed. by R. Menary. Cambridge/London: MIT Press, 325–353.

Sanz, C. M., J. Call, and C. Boesch, eds. (2013). *Tool Use in Animals: Cognition and Ecology*. Cambridge: Cambridge University Press.

Saygin, A. P., I. Cicekli, and V. Akman (2000). Turing Test: 50 Years Later. *Minds and Machines* 10, 463–518.

Sayre, K. M. (1983). Some Untoward Consequences of Dretske's "Causal Theory" of Information. *Behavioral and Brain Sciences* 6, 78–79.

Scheler, M. (1928). *Die Stellung des Menschen im Kosmos*. Darmstadt: Otto Reichl Verlag.

Searle, J. R. (1980). Minds, Brains, and Programs. *The Behavioral and Brain Sciences* 3, 417–457.

———. (1983). *Intentionality: An Essay in the Philosophy of Mind*. Cambridge: Cambridge University Press.

Sedgwick, D. (2013). *Continental Watches Eyeballs to Limit Distraction*. URL: http://www.autonews.com/apps/pbcs.dll/article?AID=/20130207/OEM10/130209894/continental-watches-eyeballs-to-limit-distraction (visited on 02/07/2013).

Sellars, R. W. (1922). *Evolutionary Naturalism*. Chicago: Open Court.

Shah, J. and C. Breazeal (2010). An Empirical Analysis of Team Coordination Behaviors and Action Planning with Application to Human – Robot Teaming. *Human Factors: The Journal of the Human Factors and Ergonomics Society* 52.2, 234–245. DOI: 10.1177/0018720809350882.

Shannon, C. E. and W. Weaver (1949). *The Mathematical Theory of Communication*. Urbana: University of Illinois Press.

Sharples, S., G. Burnett, and S. Cobb (2014). Sickness in Virtual Reality. In: *Advances in Virtual Reality and Anxiety Disorders*. Ed. by B. K. Wiederhold and S. Bouchard. Series in Anxiety and Related Disorders. New York: Springer, 35–62. ISBN: 978-1-4899-8022-9. DOI: 10.1007/978-1-4899-8023-6_3.

Shumaker, R. W., K. R. Walkup, and B. B. Beck (2011). *Animal Tool Behavior: The Use and Manufacture of Tools by Animals*. 2nd ed. Baltimore: Johns Hopkins University Press.

Simon, H. A. (1969). *The Sciences of the Artificial*. 2nd ed. Cambridge/London: MIT Press.

Sixma, T. (2009). The Gorean Community in Second Life: Rules of Sexual Inspired Role-Play. *Journal of Virtual Worlds Research* 1.3, 3–18.

Skyrms, B. (2010). *Signals: Evolution, Learning, and Information*. Oxford: Oxford University Press.

Smart, J. (1963). *Philosophy and Scientific Realism*. London: Routledge & Kegan Paul.

Sober, E. (1998). Six Sayings about Adaptationism. In: *The Philosophy of Biology*. Ed. by D. L. Hull and M. Ruse. Oxford: Oxford University Press, 72–86.

Souza e Silva, A. de. (2009). Hybrid Reality and Location-Based Gaming: Redefining Mobility and Game Spaces in Urban Environments. *Simulation & Gaming* 40.3, 404–424.

Souza e Silva, A. de and D. M. Sutko, eds. (2009). *Digital Cityscapes: Merging Digital and Urban Playspaces*. Frankfurt: Peter Lang.

Sprevak, M. (2009). Extended Cognition and Functionalism. *Journal of Philosophy* 106.9, 503–527.

Stanovsky, D. (2004). Virtual Reality. In: *The Blackwell Guide to the Philosophy of Computing and Information*. Ed. by L. Floridi. Oxford: Blackwell, 167–177.

Steels, L. and R. Brooks (1995). *The Artificial Life Route to Artificial Intelligence: Building Embodied, Situated Agents*. Hillsdale: Erlbaum.

Stenros, J., A. Waern, and M. Montola (2012). Studying the Elusive Experience in Pervasive Games. *Simulation & Gaming* 43.3, 339–355.

Sterelny, K. (2003). *Thought in a Hostile World*. Oxford: Blackwell.

———. (2004). Externalism, Epistemic Artefacts and the Extended Mind. In: *The Externalist Challenge: New Studies on Cognition and Intentionality*. Ed. by R. Schantz. Berlin/New York: de Gruyter, 239–254.

———. (2010). Minds: Extended or Scaffolded? *Phenomenology and the Cognitive Sciences* 9, 465–481.

Sterelny, K. and P. E. Griffiths (1999). *Sex and Death: An Introduction to Philosophy of Biology*. Chicago: University of Chicago Press.

Sterrett, S. S. (2000). Turing's Two Tests for Intelligence. *Minds and Machines* 10, 541–559.

Stoljar, D. (2009). Physicalism. In: *The Stanford Encyclopedia of Philosophy*. Ed. by E. N. Zalta. Spring. Stanford: The Metaphysics Research Lab, html. URL: http://plato.stanford.edu/archives/spr2009/entries/physicalism/.

Suchman, L. A. (2008). Subject Objects. Paper presented at 4S/EASST Conference 2008, Presidential Plenary: 'Acting' with 'Innovative Technologies'. Rotterdam.

Sutton, J. (2010). Exograms and Interdisciplinarity: History, the Extended Mind, and the Civilizing Process. In: *The Extended Mind*. Ed. by R. Menary. Cambridge/London: MIT Press, 189–225.

Teschke, I., C.A.F. Wascher, M. F. Scriba, A.M.P. von Bayern, V. Huml, B. Siemers, and S. Tebbich. (2013). Did Tool-Use Evolve with Enhanced Physical Cognitive Abilities? *Philosophical Transactions of the Royal Society of London, B* 368, 20120418.

Tesla Motors (2016). *A Tragic Loss*. URL: https://www.tesla.com/de_DE/blog/tragic-loss?redirect=no (visited on 06/30/2016).

Tomasello, M. (1999). *The Cultural Origins of Human Cognition*. Cambridge: Harvard University Press.

———. (2014). *A Natural History of Human Thinking*. Cambridge: Harvard University Press.

Turing, A. M. (1936). On Computable Numbers, with an Application to the Entscheidungsproblem. *Proceedings of the London Mathematical Society* s2–42, 230–265.

———. (1950). Computing Machinery and Intelligence. *Mind* 59, 433–460.

Turvey, M. (1992). Affordances and Prospective Control: An Outline of an Ontology. Ecological *Psychology* 4.3, 173–187.

———. (2013). Affordance: Towards an Ontology for all Organisms. Paper presented at "What Affordance Affords" workshop, TU Darmstadt, Germany, on 25 November, 2013.

Turvey, M. and R. E. Shaw (1979). The Primacy of Perceiving: An Ecological Reformulation of Perception for Understanding Memory. In: *Perspectives on Memory Research: Essays in Honor of Uppsala University's 500th Anniversary*. Ed. by L.-G. Nilsson. Hillsdale: Lawrence Erlbaum Associates, 167–222.

Turvey, M., R. E. Shaw, E. S. Reed, and W. M. Mace (1981). Ecological Laws of Perceiving and Acting: In Reply to Fodor and Pylyshyn. *Cognition* 9, 237–304.

Uexküll, J. von. (1956). *Streifzüge durch die Umwelten von Tieren und Menschen/ Bedeutungslehre*. Rowohlts Deutsche Enzyklopädie. Hamburg: Rowohlt.

Warren, W. J. J. (1984). Perceiving Affordances: Visual Guidance of Stair Climbing. *Journal of Experimental Psychology* 10.5, 683–703.

Weaver, W. (1949). The Mathematics of Communication. *Scientific American* 181.1, 11–15.
Weiser, M. (1991). The Computer for the 21st Century. *Scientific American* 265.9, 94–104.
Weizenbaum, J. (1976). *Computer Power and Human Reason.* San Francisco: Freeman & Co.
West, M. J. and A. P. King (1987). Settling Nature and Nurture into an Ontogenetic Niche. *Developmental Psychobiology* 20.5, 549–562.
What Is a Machine? (1872). *Scientific American* 26.3, 39.
Wheeler, M. (2004). Is Language the Ultimate Artefact? *Language Sciences* 26, 693–715.
———. (2010). In Defense of Extended Functionalism. In: *The Extended Mind.* Ed. by R. Menary. Cambridge/London: MIT Press, 245–270.
Wheeler, M. and A. Clark (1999). Genic Representation: Reconciling Content and Causal Complexity. *British Journal for the Philosophy of Science* 50, 103–135.
———. (2008). Culture, Embodiment and Genes: Unravelling the Triple Helix. *Philosophical Transactions of the Royal Society, B* 363.1509, 3563–3575.
Whitby, B. (1996). The Turing Test: AI's Biggest Blind Alley? In: *Machines and Thought.* Ed. by P. Millican and A. Clark. Vol. 1. The Legacy of Alan Turing. Oxford: Clarendon Press, 53–62.
Wieser, E., P. Mittendorfer, and G. Cheng (2011). Accelerometer based robotic Joint Orientation Estimation. In: *11th IEEE-RAS International Conference on Humanoid Robots.* New York: IEEE, 67–74.
Williams, G. C. (1966). *Adaptation and Natural Selection: A Critique of Some Current Evolutionary Thought.* Princeton: Princeton University Press.
Wilson, R. A. (1994). Wide Computationalism. *Mind* 103.411, 351–372.
Wilson, R. A. and A. Clark (2009). How to Situate Cognition: Letting Nature Take Its Course. In: *The Cambridge Handbook of Situated Cognition.* Ed. by P. Robbins and M. Aydede. Cambridge: Cambridge University Press, 55–77.
Wit, M. M. de, J. van der Kamp, and R. Withagen (2015). Visual Illusions and Direct Perception: Elaborating on Gibson's Insights. *New Ideas in Psychology* 36, 1–9.
Withagen, R. and A. Chemero (2009). Naturalizing Perception: Developing the Gibsonian Approach to Perception along Evolutionary Lines. *Theory & Psychology* 19.3, 363–389.
Wittgenstein, L. (1933). *Tractatus Logico-Philosophicus.* Ed. by C. K. Ogden. 2nd ed. London: Routledge and Kegan Paul.
———. (1969). *Philosophische Grammatik.* Ed. by R. Rhees. Frankfurt: Suhrkamp.
———. (1984). *Bemerkungen über die Philosophie der Psychologie.* Vol. 7. Werkausgabe. Frankfurt: Suhrkamp.
Yee, N., J. N. Bailenson, M. Urbanek, F. Chang, and D. Merget (2007). The Unbearable Lightness of Being Digital: The Persistence of Nonverbal Social Norms in Online Virtual Environments. *Cyberpsychology and Behavior* 10.1, 115–121.
Zhou, F., H. B.-L. Duh, and M. Billinghurst (2008). Trends in Augmented Reality Tracking, Interaction and Display: A Review of Ten Years of ISMAR. In: *Proceedings of ISMAR 2008.* Washington, DC: IEEE Computer Society, 193–202.
Zhu, Q. and G. P. Bingham (2011). Human Readiness to Throw: The Size – Weight Illusion Is Not an Illusion When Picking the Best Objects to Throw. *Evolution and Human Behavior* 32, 288–293.

Author Index

Subject Index